**DEPARTMENT OF THE ARMY**
**U.S. Army Corps of Engineers**
**Washington, DC 20314-1000**

CECW-CE

EM 1110-2-1100
(Change 2)

Manual
No. 1110-2-1100

1 April 2008

**Engineering and Design**
**COASTAL ENGINEERING MANUAL**

**1. Purpose.** The purpose of the *Coastal Engineering Manual* (CEM) is to provide a comprehensive technical coastal engineering document. It includes the basic principles of coastal processes, methods for computing coastal planning and design parameters, and guidance on how to formulate and conduct studies in support of coastal flooding, shore protection, and navigation projects. This Change 2 to EM 1110-2-1100, 1 April 2008, includes the following changes and updates:

    *a.* Part I-1. References were checked and some were deleted (Engineer Manuals that are no longer in the USACE inventory).

    *b.* Part I-4. Minor changes were made in the text to better reflect the contents of subsequent parts of the CEM.

    *c.* Part II-1. Figure II-1-9 has been revised; Equations II-1-128, II-1-160, and II-1-161 have been corrected.

    *d.* Part II-2. Equations II-2-4, II-2-5, and II-2-32 have been corrected along with other errors reported by various users.

    *e.* Part II-5. References were checked and some were deleted (Engineer Manuals that are no longer in the USACE inventory).

    *f.* Part II-6. The value of "e" used in Eq. II-6-28 has been corrected.

    *g.* Part II-7. The table of contents was corrected. A new section, II-7-11, Note to Users, Vessel Buoyancy, was added at the end of the chapter.

    *h.* Part III-3. Corrections have been made to format and spelling. Different plots were added to Figures III-3-24 and III-3-26.

    *i.* Part IV-1. Corrections have been made to references.

    *j.* Part V-1. Citation of an Engineer Regulation has been corrected.

    *k.* Part V-2. Citation of references has been changed, web pages with sources of wind and wave data have been added. Some minor text changes have also been made.

    *l.* Part V-3. Citations of unpublished reports or personal communications have been deleted, and links to other figures or parts of the CEM have been checked and corrected.

    *m.* Part V-4. Minor text changes, corrections to references and Figure V-4-1.

    *n.* Part V-5. Links to other parts of the CEM that were planned but never written have been deleted.

**2. Applicability.** This manual applies to all HQUSACE elements and all USACE commands having Civil Works and military design responsibilities.

**3. Discussion.** The CEM is divided into five parts in two major subdivisions: science-based and engineering-based. The first four parts of the CEM and Appendix A compose the science-based subdivision:

Part I, "Introduction"
Part II, "Coastal Hydrodynamics"
Part III, "Coastal Sediment Processes"
Part IV, "Coastal Geology"
Appendix A, "Glossary"

The engineering-based subdivision is oriented toward a project-type approach, Part V, "Coastal Project Planning and Design."

**4. Distribution Statement.** Approved for public release, distribution unlimited.

**5. Note to Users.** Revised chapters are dated 1 April 2008. Readers need to download the entire new chapters and discard earlier versions in their possession.

FOR THE COMMANDER:

STEPHEN L. HILL
Colonel, Corps of Engineers
Chief of Staff

Chapter 1
COASTAL TERMINOLOGY AND
GEOLOGIC ENVIRONMENTS

EM 1110-2-1100
(Part IV)
1 August 2008 (Change 2)

## Table of Contents

# List of Tables

# List of Figures

Page

## Chapter IV-1
## Coastal Terminology and Geologic Environments

### IV-1-1. Background

*a.* Since man has ventured to the sea, he has been fascinated by the endless variety of landforms and biological habitats that occur at the coast. With the exception of high altitude alpine, a full spectrum of environments is found around the world's coastlines. These range from icy arctic shores to rocky faulted coasts to temperate sandy barriers to tropical mangrove thickets, with a myriad of intermediate and mixed forms. This part of the *Coastal Engineering Manual* (CEM) concentrates on the geology of the coastal zone. This broad subject encompasses both the geomorphology (the shape and form) of the landforms and the nature of the ancient strata that underlie or outcrop in the region. The forces that shape, and are shaped by, the coast are part of the overall picture, although here geology merges with the other earth sciences of meteorology and oceanography.

*b.* This and the two chapters that follow have ambitious goals:

(1) To review overall geological, environmental, and climatological settings of the world's coasts.

(2) To describe particular shore types in detail and provide examples.

(3) To explain how shore types are created by and interact with the forces of waves, currents, and weather (sometimes known as "morphodynamics").

The emphasis here is on features and landforms that range in size from centimeters to kilometers and are formed or modified over time scales of minutes to millennia (Figure IV-1-1).

*c.* Another subject of crucial importance to coastal researchers is biology. The biological environment is partly established by the geological setting. Conversely, biology affects coastal geology in many ways:

(1) Coral reefs and mangroves have created large stretches of coastline.

(2) Cliff erosion is accelerated by the chemical solution and mechanical abrasion caused by some organisms.

(3) Lagoons and estuaries slowly fill with the by-products of plants and the sediment they trap, forming wetlands.

These topics are briefly reviewed in this text, but details of the flora and fauna that inhabit the coast are not covered here.

*d.* Field methods and data analysis procedures applicable to geological field studies at the coast are not reviewed in this part of the CEM. Field methods are constantly evolving and changing as new instruments and technology are developed. Coastal monitoring methods were reviewed in Gorman, Morang, and Larson (1998); Larson, Morang, and Gorman (1997); and Morang, Larson, and Gorman (1997a,b). Readers who plan to conduct field studies should contact surveyors or contractors familiar with the latest technology and should review trade journals that discuss oceanographic, geographic information systems (GIS), remote sensing, and surveying advances.

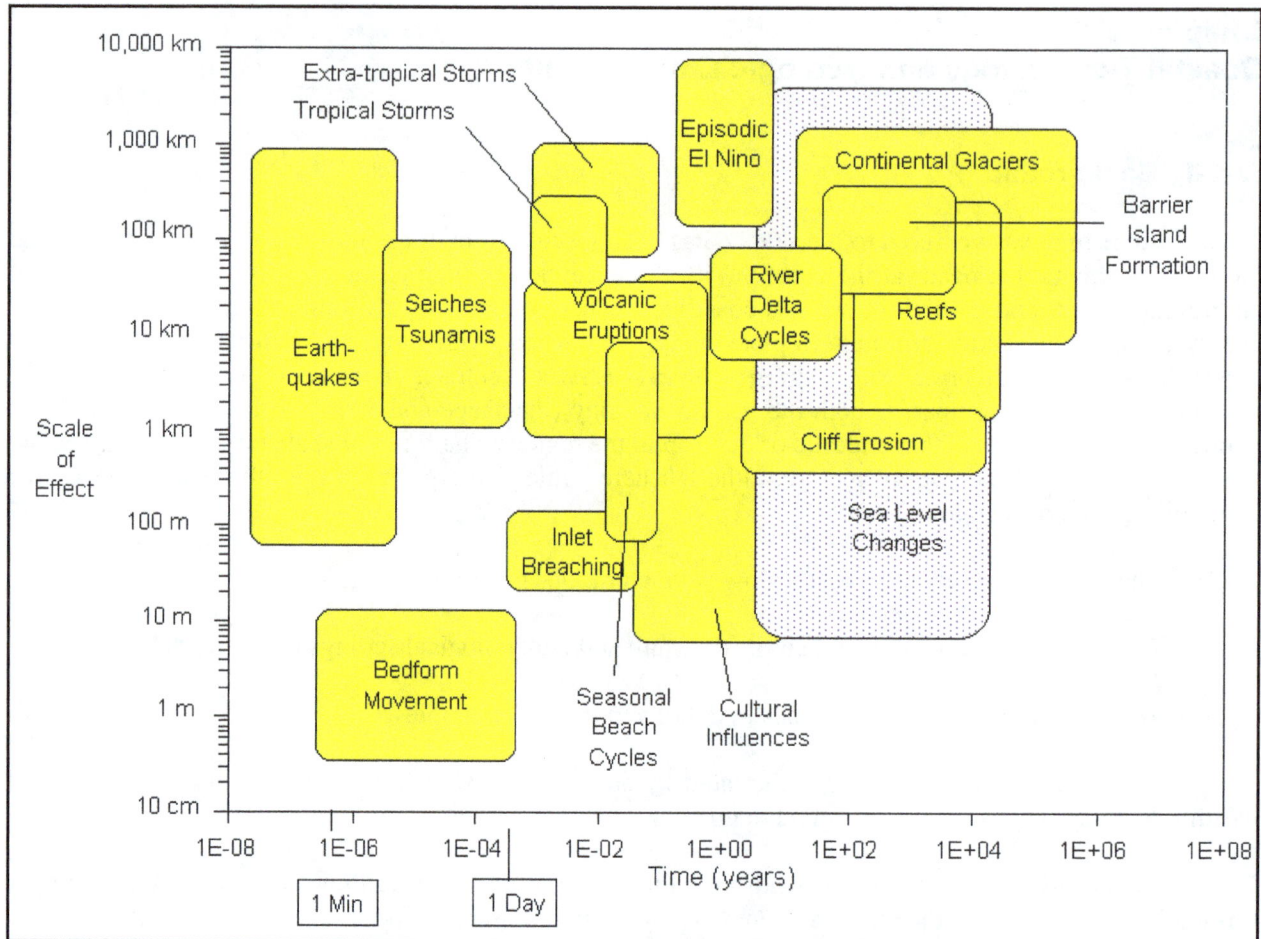

**Figure IV-1-1.  Temporal and spatial scales of geologic and oceanographic phenomena**

## IV-1-2.  Coastal Zone Definitions and Subdivisions

*a.  Introduction.*

(1)  Many coastal zone features and subdivisions are difficult to define because temporal variability or gradational changes between features obscure precise boundaries.  In addition, nomenclature is not standardized, and various authors describe the same features using different names.  If the same name is used, the intended boundaries may differ greatly.  This ambiguity is especially evident in the terminology and zonation of shore and littoral areas.  In the absence of a widely accepted standard nomenclature, coastal researchers would do well to accompany reports and publications with diagrams and definitions to ensure that readers will fully understand the authors' use of terms.

(2)  The following subparagraphs present coastal zone definitions and subdivisions based largely, but not exclusively, on geological criteria.  They do not necessarily coincide with other geological-based zonations or those established by other disciplines.  It should be borne in mind that coastal zone geology varies greatly from place to place, and the zonations discussed below do not fit all regions of the world.  For example, coral atolls are without a coast, shoreface, or continental shelf in the sense defined here.  The Great Lakes and other inland water bodies have coasts and shorefaces but no continental shelves.  Thus, while divisions and categories are helpful in describing coastal geology, flexibility and good descriptive text and illustrations are always necessary for adequate description of a given region or study site.

*b. Coastal zone.* In this manual, *coastal zone* is defined as the transition zone where the land meets water, the region that is directly influenced by marine or lacustrine hydrodynamic processes. The coastal zone extends offshore to the continental shelf break and onshore to the first major change in topography above the reach of major storm waves. Although this discussion excludes upland rivers, river mouth deltas, where morphology and structure are a result of the dynamic interplay of marine and riverine forces, are included. The coastal zone is divided into four subzones (Figures IV-1-2 and IV-1-3):

(1) Coast.

(2) Shore.

(3) Shoreface.

(4) Continental shelf.

*c. Coast.* The *coast* is a strip of land of indefinite width that extends from the coastline inland as far as the first major change in topography. Cliffs, frontal dunes, or a line of permanent vegetation usually mark this inland boundary. On barrier coasts, the distinctive back-barrier lagoon/marsh/tidal creek complex is considered part of the coast. It is difficult to define the landward limit of the coast on large deltas like the Mississippi, but the area experiencing regular tidal exchange can serve as a practical limit (in this context, New Orleans would be considered "coastal"). The seaward boundary of the coast, the *coastline*, is the maximum reach of storm waves. On shorelines with plunging cliffs, the coast and coastline are one and the same. It is difficult to decide if a seawall constitutes a coast; the inland limit might better be defined at a natural topographic change.

*d. Shore.* The *shore* extends from the low-water line to the normal landward limit of storm wave effects, i.e., the coastline. Where beaches occur, the shore can be divided into two zones: *backshore* (or berm) and *foreshore* (or beach face). The foreshore extends from the low-water line to the limit of wave uprush at high tide. The backshore is horizontal while the foreshore slopes seaward. This distinctive change in slope, which marks the juncture of the foreshore and backshore, is called the beach or berm crest.

*e. Shoreface.* The *shoreface* is the seaward-dipping zone that extends from the low-water line offshore to a gradual change to a flatter slope denoting the beginning of the continental shelf. The continental shelf transition is the *toe of the shoreface*. Its location can only be approximately marked due to the gradual slope change. Although the shoreface is a common feature, it is not found in all coastal zones, especially along low-energy coasts or those consisting of consolidated material. The shoreface can be delineated from shore-perpendicular profile surveys or from bathymetric charts (if they contain sufficient soundings in shallow water). The shoreface, especially the upper part, is the zone of most frequent and vigorous sediment transport.

*f. Continental shelf.* The *continental shelf* is the shallow seafloor that borders most continents (Figure IV-1-3). The shelf floor extends from the toe of the shoreface to the shelf break where the steeply inclined continental slope begins. It has been common practice to subdivide the shelf into inner-, mid-, and outer zones, although there are no regularly occurring geomorphic features on most shelves that suggest a basis for these subdivisions. Although the term *inner shelf* has been widely used, it is seldom qualified beyond arbitrary depth or distance boundaries. Site-specific shelf zonation can be based on project

## a. Typical Beach Profile

Coastline

Low Water Line

Toe of Shoreface
(Approx. Position)

Coast

Shore

Shoreface

Upland

Beach

Nearshore

Offshore

Backshore

Foreshore

Dune
Crest

Swash
Zone *

Surf *
Zone

Shore *
Breaker
Zone

Surf Zone *

Bar *
Breaker
Zone

Transition
Zone

Dune

Storm
Berm

Berm
Crests

Dune
Scarp

Berm

Berm
Scarp

MHW

* — Location and Width Vary
as Wave Conditions Change

Low
Tide
Terrace

Step

MLW

Trough

Bar

Trough and Bar
Position Variable

**b. Typical Bluff Profile**

Bluff

Berm
Crest

Berm

MHW

MLW

Step

Bar

Trough

**c. Typical Overwash Profile**

Overwash
Fan

Berm
Crest

MHW

Low
Tide
Terrace

Runnel

Ridge

MLW

Step

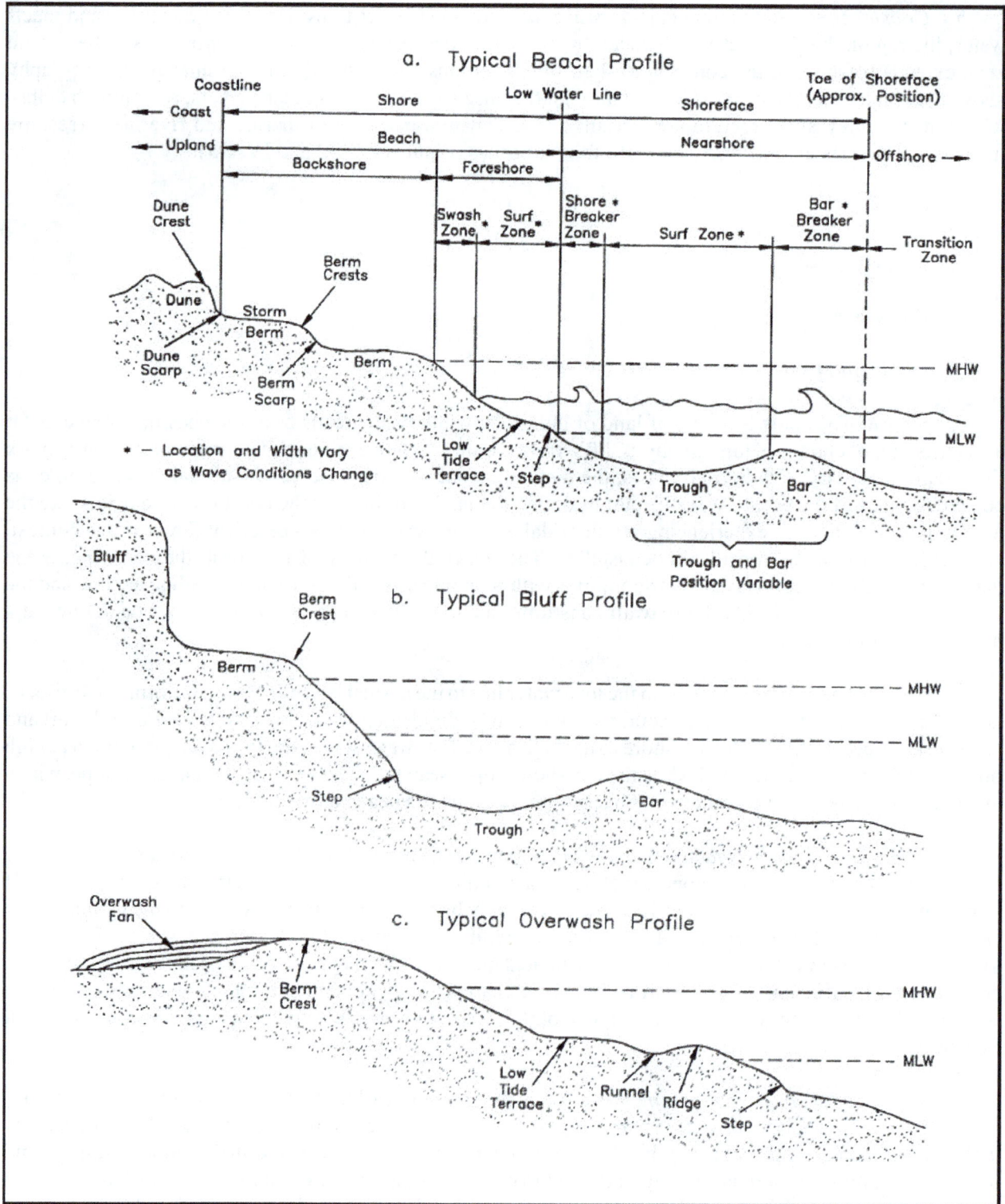

**Figure IV-1-2. Definition of terms and features describing the coastal zone**

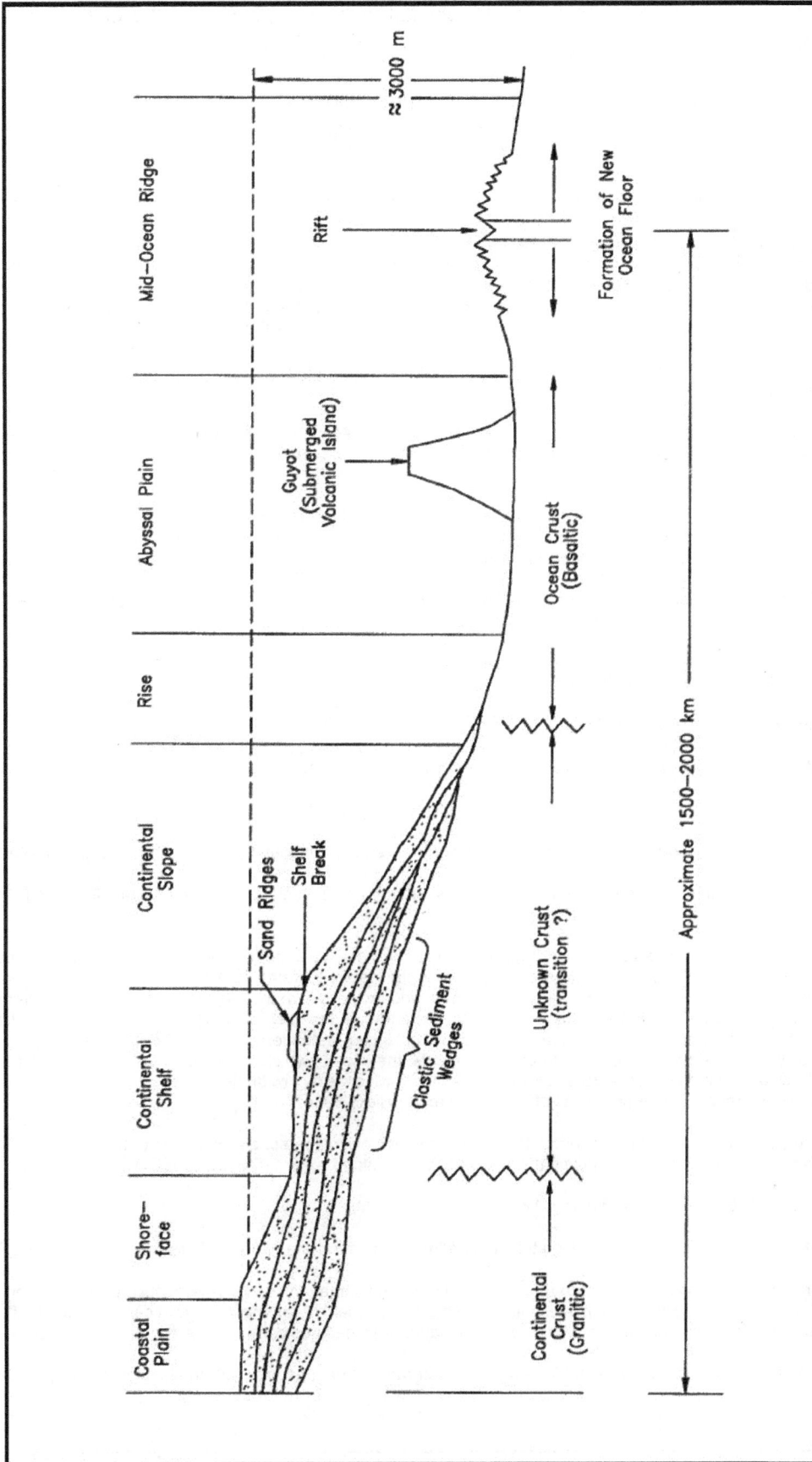

Figure IV-1-3. Continental shelf and ocean floor along the trailing-edge of a continent (i.e., representative of the U.S. Atlantic Ocean coast). Figure not to scale, great vertical exaggeration)

requirements and local geologic conditions. Some coastal areas (e.g., bays and the Great Lakes) do not extend out to a continental shelf.

   *g. Shoreline definitions.* Common coastal geomorphic features are defined in Table IV-1-1 below. These have been adapted from the *National Shoreline Data Standard,* a draft standard prepared by the National Oceanic and Atmospheric Administration (NOAA).

---

**Table IV-1-1**
**Definitions of Common Coastal Geomorphic Features**

**Apparent shoreline**   Line drawn on a map or chart in lieu of a mean high-water line or the mean water level line in areas where either may be obscured by marsh, mangrove, cypress, or other type of marine vegetation. This line represents the intersection of the appropriate datum on the outer limits of vegetation and appears to the navigator as the shoreline (Ellis 1978).

**Backshore**   That part of the beach that is usually dry, being reached only by the highest tides, and by extension, a narrow strip of relatively flat coast bordering the sea (Ellis 1978).

**Bank**   Edge of a cut or fill; the margin of the watercourse; an elevation of the seafloor located on a continental shelf or an island shelf and over which the depth of water is relatively shallow but sufficient for safe surface navigation (reefs or shoals, dangerous to surface navigation may arise above the general depths of a bank) (Ellis 1978).

**Beach (or seabeach)**   Zone of unconsolidated material that extends landward from the low-water line to the place where there are marked changes in material or physiographic form, or to the line of permanent vegetation (usually the effective limit of storm waves). A beach includes foreshore and backshore (Ellis 1978).

**Beach berm**   Nearly horizontal portion of the beach or backshore formed by the deposit of materials by wave action.  Some beaches have no berms, others have one or several (Ellis 1978).

**Berm**   Nearly horizontal portion of a beach or backshore having an abrupt fall and formed by wave deposition of material and marking the limit of ordinary high tides (Ellis 1978).

**Berm crest**   Seaward limit of a berm (Ellis 1978).

**Bluff**   A cliff or headland with an almost perpendicular face (International Hydrographic Bureau 1990).

**Bottom lands**   Land below navigable freshwater bodies (Coastal States Organization 1997).

**Cliff**   Land rising abruptly for a considerable distance above the water or surrounding land (Hydrographic Dictionary 1990).

**Coast**   General region of indefinite width that extends from the sea inland to the first major change in terrain features (Ellis 1978).

**Coastal zone (legal definition for coastal zone management)**   The term coastal zone means the coastal waters (including the lands therein and thereunder) and the adjacent shorelands (including the waters therein and thereunder), strongly influenced by each and in proximity to the shorelines of the several coastal states, and includes islands, transitional and inter-tidal areas, salt marshes, wetlands, and beaches. The zone extends, in Great Lakes waters, to the international boundary between the Unites States and Canada and in other areas seaward to the outer limit of the United States territorial sea. The zone extends inland from the shorelines only to the extent necessary to control shorelands, the uses of which have a direct and significant impact on the coastal waters. Excluded from the coastal zone are lands the use of which is by law subject solely to the discretion of or which is held in trust by the Federal Government, its officers, or agents (Hicks 1984).

**Coast line (According to Public Law 31)**   Defined as the line of ordinary low water along that portion of the coast that is in direct contact with the open sea and the line marking the seaward limit of inland waters (Shalowitz 1964).

**Coastline**   Same as shoreline (see coast line) (Hicks 1984).

**Dry sand beach**   Sandy area between the mean high tide line and the vegetation line (Coastal States Organization 1997).

**Estuary**   An embayment of the coast in which fresh river water entering at its head mixes with the relatively saline ocean water. When tidal action is the dominant mixing agent it is usually termed a tidal estuary. Also, the lower reaches and mouth of a river emptying directly into the sea where tidal mixing takes place. The latter is sometimes called a river estuary (Hicks 1984).

**Foreshore**   That part of shore which lies between high and low water mark at ordinary tide (International Hydrographic Bureau 1990).

---

**Freshwaters** Waters that do not ebb and flow with the tide. The determinative factor is that the water body does not ebb and flow with the tide, not the salt content of the water (Coastal States Organization 1997).

**High-water line** A generalized term associated with the tidal plane of high water but not with a specific phase of high water (for example, higher high water, lower high water) (Shalowitz 1964).

**High-water mark** A line or mark left upon tide flats, beach, or alongshore objects indicating the elevation of the intrusion of high water. The mark may be a line of oil or scum along shore objects, or a more or less continuous deposit of fine shell or debris on the foreshore or berm. This mark is physical evidence of the general height reached by wave runup at recent high waters. It should not be confused with the mean high water line or mean higher high water line (Hicks 1984).

**Inshore** In beach terminology, the zone of variable width between the shoreface and the seaward limit of the breaker zone (Ellis 1978).

**Intertidal zone (technical definition)** The zone between the mean higher high water and mean lower low water lines (Hicks 1984).

**Island** A piece of land completely surrounded by water (International Hydrographic Bureau 1990).

**Ledge** A shelf -like projection, on the side of a rock or mountain. A rocky formation continuous with and fringing the shore (International Hydrographic Bureau 1990).

**Levee** Artificial bank confining a stream channel or limiting adjacent areas subject to flooding; an embankment bordering a submarine canyon or channel, usually occurring along the outer edge of a curve (Ellis 1978).

**Littoral** Pertaining to the shore, especially of the sea; a coastal region. Used extensively with "riparian" (Shalowitz 1964).

**Shorelands** General term including tidelands and navigable freshwater shores below the ordinary high-water mark (Coastal States Organization 1997).

**Shoreline** The line of contact between the land and a body of water. On Coast and Geodetic Survey nautical charts and surveys, the shoreline approximates the mean high-water line. In Coast Survey usage, the term is considered synonymous with coastline (Shalowitz 1962).

**Submerged lands** Lands covered by water at any stage of the tide, as distinguished from tidelands, which are attached to the mainland or an island and cover and uncover with the tide. Tidelands presuppose a high-water line as the upper boundary; submerged lands do not (Shalowitz 1962).

**Tidal estuary** See estuary (Hicks 1984).

**Tidelands** The land that is covered and uncovered by the daily rise and fall of the tide. More specifically, it is the zone between the mean high waterline and the mean low waterline along the coast, and is commonly known as the "shore" or "beach." Referred to in legal decisions as between the ordinary high-water mark and ordinary low-water mark. Tidelands presuppose a high-water line as the upper boundary (Shalowitz 1964).

**Tidewaters** Waters subject to the rise and fall of the tide. Sometimes used synonymously with tidelands, but tidewaters are better limited to areas always covered with water. The amount of tide is immaterial (Shalowitz 1964).

**Upland** Land above mean high water mark and subject to private ownership, as distinguished from tidelands, ownership of which is prima facie in the state but also subject to divestment under state statutes (Shalowitz 1964).

**Waterline** Juncture of land and sea. This line fluctuates, changing with the tide or other fluctuations in the water (Ellis 1978).

**Wet sand beach** Area between the ordinary high tide and the ordinary low tide lines (Coastal States Organization 1997).

Source: National Oceanic and Atmospheric Administration (NOAA) 1998. National Shoreline Data Standard, Progress Report and Preliminary Draft Standard, NOAA Office of Coast Survey, Silver Spring, MD.

## IV-1-3. Geologic Time and Definitions

*a. Geologic fossil record.* Geologists have subdivided geologic time into *eras, periods,* and *epochs* (Figure IV-1-4). Pioneering geologists of the 1800's based the zonations on the fossil record when they discovered that fossils in various rock formations appeared and disappeared at distinct horizons, thus providing a means of comparing and correlating the relative age of rock bodies from widely separated

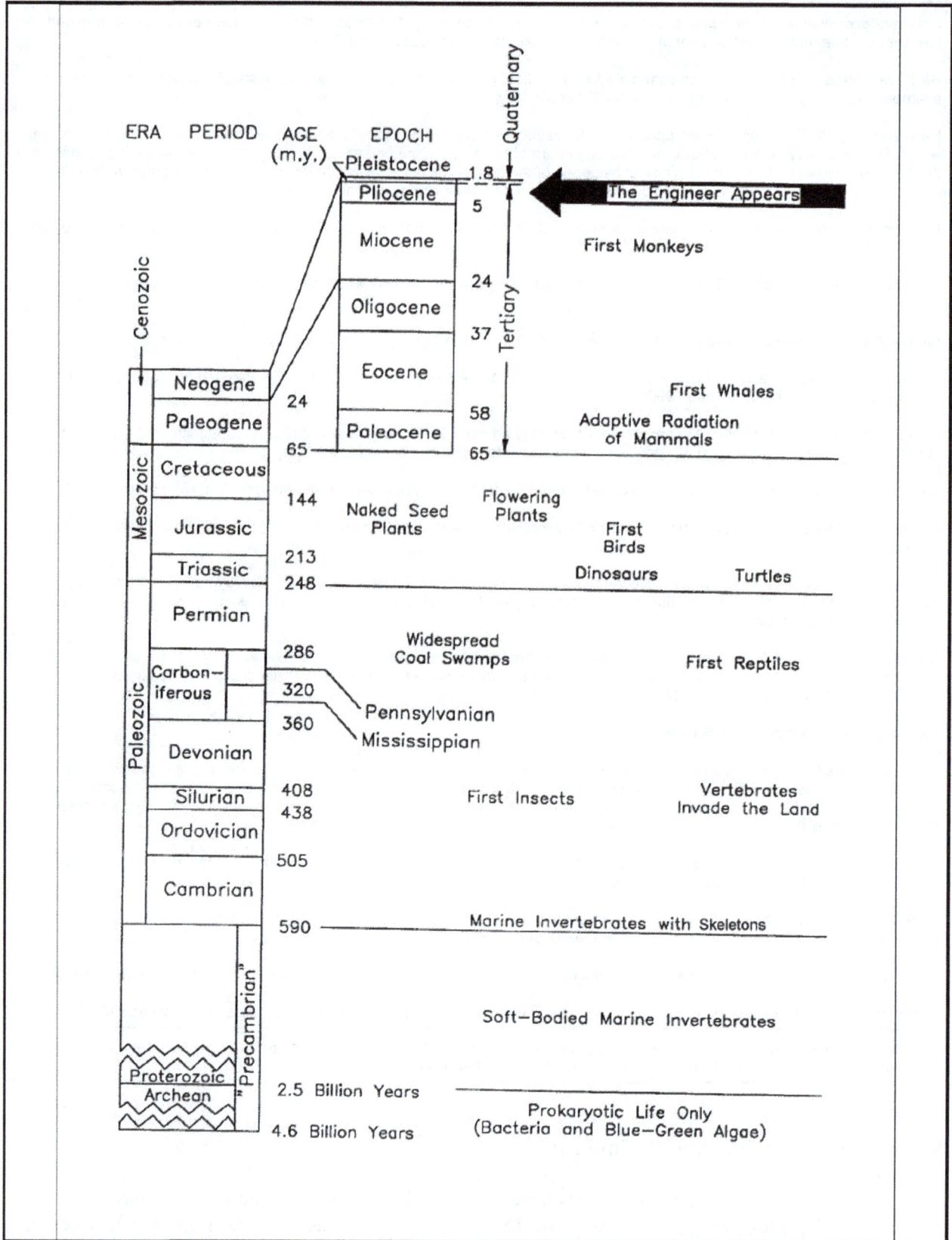

**Figure IV-1-4. Geologic time scale. Chronological ages are based on radiometric dating methods (figure adapted from Stanley (1986))**

locations. For example, the boundary between the *Mesozoic* ("interval of middle life") and the *Cenozoic* ("interval of modern life") eras is marked by the disappearance of hundreds of species, including the dinosaurs, and the appearance or sudden proliferation of many new species (Stanley 1986). The fossil time scale was relative, meaning that geologists could compare rock units but could not assign absolute ages in years. It was not until the mid-20th century that scientists could measure the absolute age of units by radiometric dating. The geologic times listed in Figure IV-1-4, in millions of years, are best estimates based on radiometric dates.

    *b. Geologic time considerations for coastal engineering.* The epochs of most concern to coastal engineers and geologists are the *Pleistocene* and *Recent* (also commonly known as the *Holocene*), extending back a total of 1.8 million years before present. *Quaternary* is often used to designate the period comprising the Pleistocene and Recent Epochs.

    (1) The Pleistocene Epoch was marked by pronounced climatic fluctuations in the Northern Hemisphere - changes that marked the modern Ice Age. The continental glaciers that periodically covered vast areas of the northern continents during this time had profound influence on the surficial geology. Many geomorphic features in North America were shaped or deposited by the ice sheets. Flint's (1971) *Glacial and Quaternary Geology* is an exhaustive study of the effects of Pleistocene ice sheets on North American geology.

    (2) The Holocene Transgression started approximately 15,000 to 18,000 years ago with the beginning of global sea level rise. Presumably, a concurrent event was the waning of the continental glaciers possibly caused by a warming climate around the world. Most of the dynamic, morphological features that we associate with the active coastal environment are Holocene in age, but the preexisting geology is often visible, as well. For example, the drumlins of Boston Harbor and the end moraine islands of southern New England (Long Island, Martha's Vineyard, and Block and Nantucket Islands) are deposits left by the Wisconsin stage glaciers (Woodsworth and Wigglesworth 1934), but barrier spits and beaches found along these shores are more recent (Holocene) features.

    (3) North American glacial stages.[1] Worldwide climatic fluctuations and multiple glacial and interglacial stages were the overwhelming Quaternary processes that shaped the surficial geomorphology and biological diversity of our world. Major fluctuations in eustatic, or worldwide, sea level accompanied the waxing and waning of the continental glaciers. Oxygen isotope analysis of deep sea sediments suggests that there were as many as nine glacial and ten interglacial events in the last 700,000 years (Kraft and Chrzastowski 1985). North American stages and approximate ages are listed in Table IV-1-2. The most recent glacial stage was the Wisconsin in North America and the Würm in Europe, during which sea level was more than 100 m below present. In northern latitude coasts, the coastal worker will often encounter geologic and geomorphic evidence of the Wisconsin glacial stage. Less evidence remains of the earlier North American stages except raised shore terraces along parts of the U.S. Atlantic and Gulf coasts (e.g., see Winkler (1977); Winkler and Howard (1977)).

## IV-1-4. Water Level Datums and Definitions

Critical in evaluating sea level information or in constructing shoreline change maps are the level and type of datum used. Because water levels are not constant over space and time, depths and elevations must be

---

[1] Stage is a time term for a major subdivision of a glacial epoch, including the glacial and interglacial events (Bates and Jackson 1984).

**Table IV-1-2**
**North American Pleistocene Glacial and Interglacial Stages**

| Age (approx. years)[1] | Glacial and Interglacial Stages | Age (approx. years)[2] |
|---|---|---|
| 12,000-Present | Recent (Holocene) | 10,000-Present |
| 150,000-12,000 | Wisconsin | 100,000-10,000 |
| 350,000-150,000 | Sangamon Interglacial | 300,000-100,000 |
| 550,000-350,000 | Illinoisan | 450,000-300,000 |
| 900,000-550,000 | Yarmouth Interglacial | 1,100,000-450,000 |
| 1,400,000-900,000 | Kansan | 1,300,000-1,100,000 |
| 1,750,000-1,400,000 | Aftonian Interglacial | 1,750,000-1,300,000 |
| > 2,000,000-1,750,000 | Nebraskan | 2,000,000-1,750,000 |
| > 2,000,000(?) | Older glaciations | |

[1] Dates based on generalized curve of ocean-water temperatures interpreted from foraminifera in deep sea cores (curve reproduced in Strahler (1981)).
[2] Dates from Young (1975) (original sources not listed).

referenced to established datums. In marine coastal areas, datums are typically based on tide elevation measurements. A glossary of tide elevation terms is presented in Part II-5-3, and water surface elevation datums are discussed in Part II-5-4.

## IV-1-5. Factors Influencing Coastal Geology

The coast is probably the most diverse and dynamic environment found anywhere on earth. Many geologic, physical, biologic, and anthropomorphic (human) factors are responsible for shaping the coast and keeping it in constant flux. Ancient geological events created, modified, and molded the rock and sediment that form the foundation of the modern coastal zone. Over time, various physical processes have acted on this preexisting geology, subsequently eroding, shaping, and modifying the landscape. These processes can be divided into two broad classes: active forces, like waves and tides, which occur constantly, and long-term forces and global changes that affect the coast over time scales of years.

   *a. Underlying geology and geomorphology.*[1] The geologic setting of a coastal site controls surficial geomorphology, sediment type and availability, and overall gradient. The geology is modified by physical processes (e.g., waves and climate), biology, and man-made activities, but the overall "look" of the coast is primarily a function of the regional lithology and tectonics. These topics are discussed in the following paragraphs.

   (1) Lithology. *Lithology* concerns the general character of rock or sediment deposits. The most critical lithologic parameters responsible for a rock's susceptibility to erosion or dissolution are the mineral composition and the degree of consolidation. Striking contrasts often occur between coasts underlain by consolidated rock and those underlain by unconsolidated material. Marine processes are most effective when acting on uncemented material, which is readily sorted, redistributed, and sculpted into forms that are in a state of dynamic equilibrium with incident energy.

---

[1] *Geomorphology* is a study of natural topographic features and patterns forming the earth's surface, including both terrestrial and subaqueous environments.

(a) Consolidated coasts. Consolidated rock consists of firm and coherent material. Coastal areas consisting of consolidated rock are typically found in hilly or mountainous terrain. Here, erosional processes are usually dominant. The degree of consolidation greatly influences the ability of a rocky coastline to resist weathering and erosion. Resistance depends on susceptibility to mechanical and chemical weathering, hardness and solubility of constituent minerals and cementation, nature and density of voids, and climatic conditions. Rock type, bedding, jointing, and orientation of the strata greatly influence the geomorphic variability of the shoreline (Figures IV-1-5, IV-2-20, and IV-2-31). For example, large portions of the shore-lines of Lakes Superior, Huron, and Ontario are rocky and prominently display the structure of the underlying geology.

- *Mechanical weathering* is the disintegration of rock without alteration of its chemical nature. Examples of mechanical weathering include fluctuations in temperature (causing repetitive thermal expansion and contraction), expansion due to crystallization from salt or ice, wetting and drying, overburden fluctuations, and biological activity.

- *Chemical weathering* is the decomposition of rock material by changes in its chemical composition. This process includes hydration and hydrolysis, oxidation and reduction, solution and carbonation, chelation, and various biochemical reactions.

(b) Unconsolidated coasts. In contrast to consolidated coasts, depositional and erosional processes domi-nate unconsolidated coasts, which are normally found on low relief coastal plains or river deltas. Commonly, shorelines have been smoothed by erosion of protruding headlands and by the deposition of barrier islands, spits, and bay mouth barriers. Along unconsolidated coasts, large amounts of sediment are usually available, and morphological changes occur rapidly. Waves and currents readily alter relict geomorphic features in this environment. Figure IV-1-6 illustrates features associated with unconsolidated depositional environments. The Atlantic and Gulf of Mexico coasts of the United States are mostly unconsolidated, depositional environments (except select locations like the rocky shores in New England).

(2) Tectonics. Forces within the earth's crust and mantle deform, destroy, and create crustal material. These tectonic activities produce structural features such as faults and folds (anticlines and synclines) (Figure IV-1-7). Tectonic movements produce large-scale uplift and subsidence of land masses. The west coast of the United States is an example of a tectonically dominated coast, in sharp contrast to the east coast, which is mostly depositional. According to Shepard's (1973) coastal classification, a fault coast is characterized by a steep land slope that continues beneath the sea surface. The most prominent feature exhibited by a fault coast is a scarp where normal faulting has recently occurred, dropping a crustal block so that it is completely submerged, leaving a higher block standing above sea level (Figure IV-1-8). Examples of fault-block coasts are found in California. Active faults such as the Inglewood-Rose Canyon structural zone outline the coast between Newport Bay and San Diego, and raised terraces backed by fossil cliffs attest to continuing tectonism (Orme 1985).

(3) Volcanic coasts. The eruption of lava and the growth of volcanoes may result in large masses of new crustal material. Conversely, volcanic explosions or collapses of existing volcanic cones can leave huge voids in the earth's surface known as calderas. When calderas and cones occur in coastal areas, the result is a coastline dominated by circular convex and concave contours (Shepard 1973). Coastlines of this sort are common on volcanic islands such as the Aleutians (Figure IV-1-9). The morphology of volcanic shores is discussed in more detail in Part IV-2.

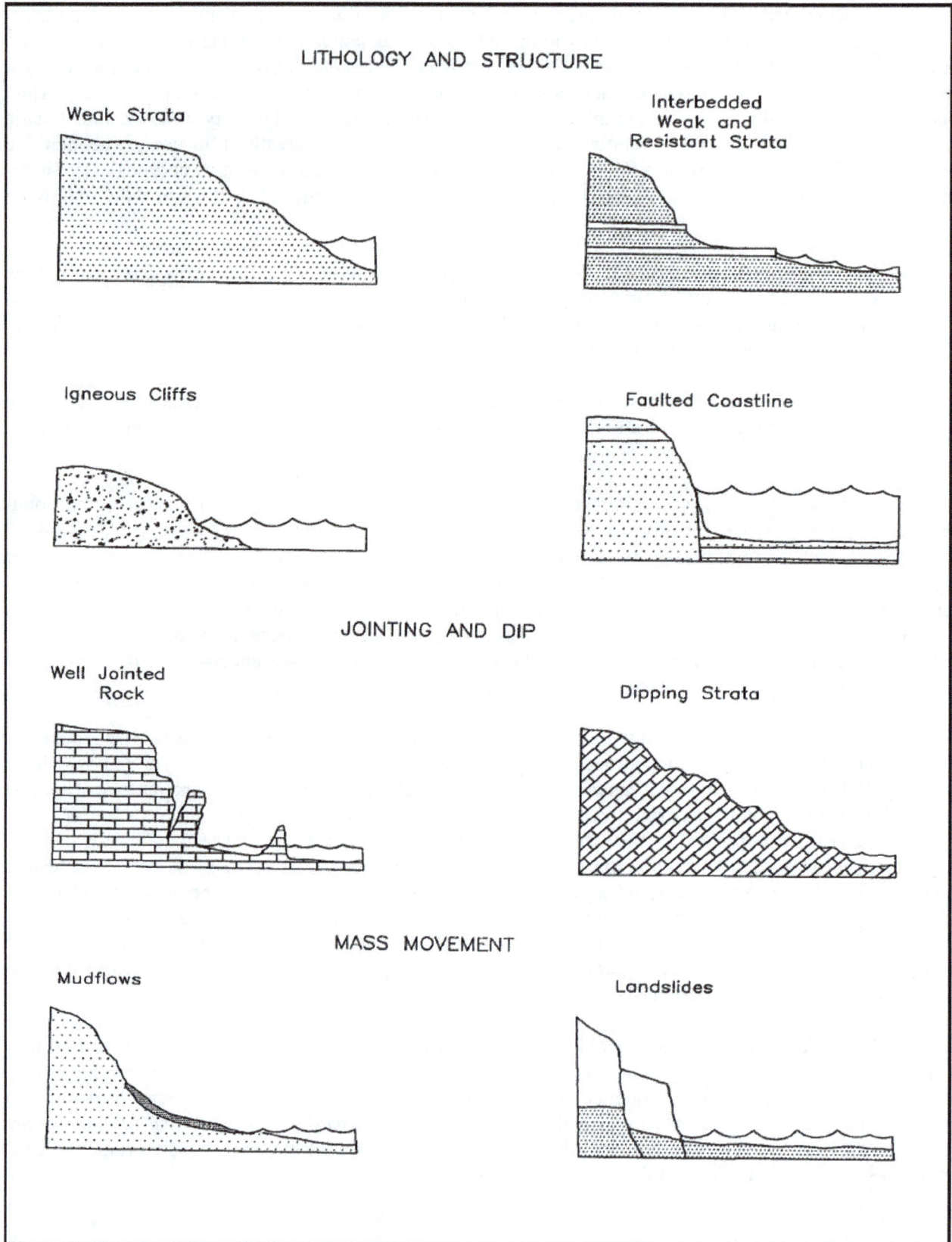

LITHOLOGY AND STRUCTURE

Weak Strata

Interbedded
Weak and
Resistant Strata

Igneous Cliffs

Faulted Coastline

JOINTING AND DIP

Well Jointed
Rock

Dipping Strata

MASS MOVEMENT

Mudflows

Landslides

Figure IV-1-5.  Cross-section views of aspects of geomorphic variability attributable to lighology, structure, and mass movement along semi-consolidated coasts (from Mossa, Meisberger, and Morang (1992))

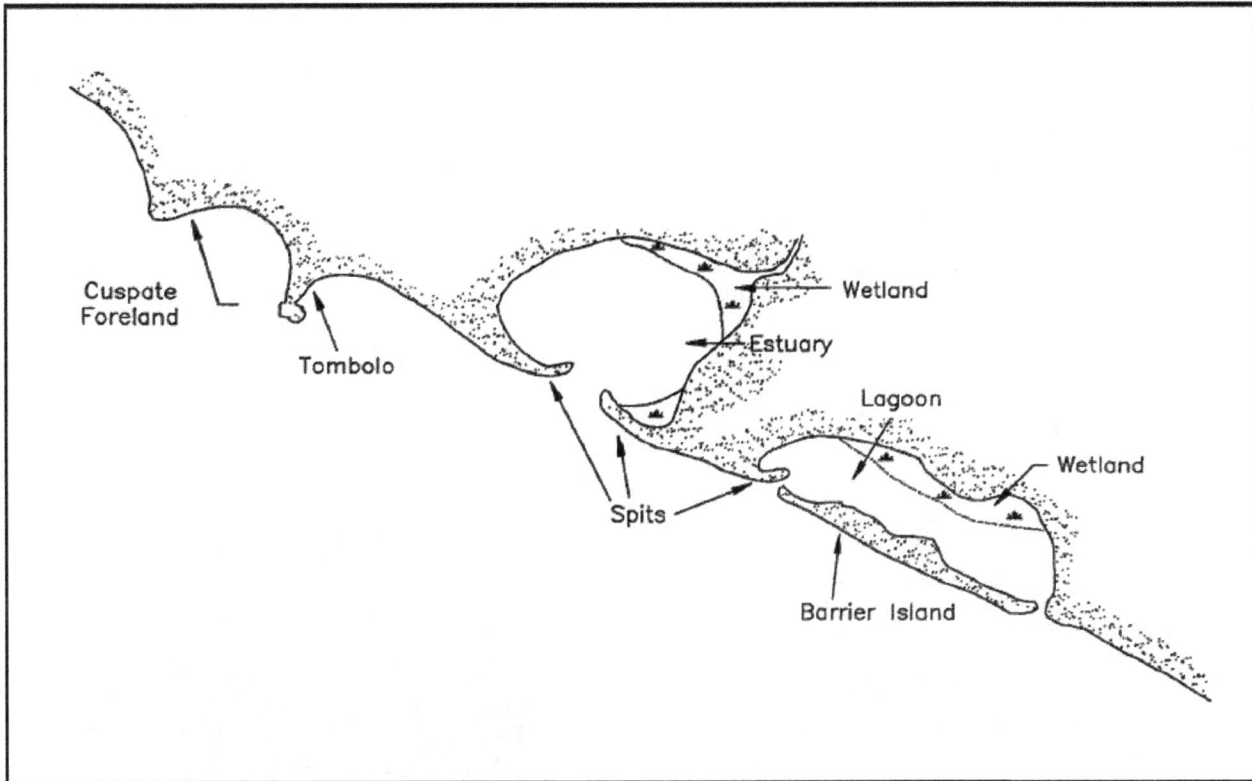

**Figure IV-1-6. Examples of features associated with depositional coastal environments. These features consist mostly of unconsolidated sediments (after Komar 1976))**

*b. High-frequency dynamic processes.* The following paragraphs briefly list processes that impart energy to the coastal zone on a continuous or, as with storms, repetitive basis. Any geological or engineering investigation of the coastal zone must consider the sources of energy that cause erosion, move sediment, deposit sediment, and rearrange or shape the preexisting topography. These processes also result in temporary changes in water levels along the coast. Long-term sea level changes are discussed in paragraph IV-1-6.

(1) Waves.

(a) Water waves (sometimes called *gravity waves*) are the dominant force driving littoral processes on open coasts. The following quotes from the *Shore Protection Manual* (1984) underscore the significance of waves in shaping coastal zone geomorphology:

Waves are the major factor in determining the geometry and composition of beaches and significantly influence the planning and design of harbors, waterways, shore protection measures, coastal structures, and other coastal works. Surface waves generally derive their energy from the winds. A significant amount of this wave energy is finally dissipated in the nearshore region and on the beaches.

Waves provide an important energy source for forming beaches; sorting bottom sediments on the shoreface; transporting bottom materials onshore, offshore, and alongshore; and for causing many of the forces to which coastal structures are subjected. An adequate understanding of the fundamental physical processes in surface wave generation and propagation must precede any attempt to

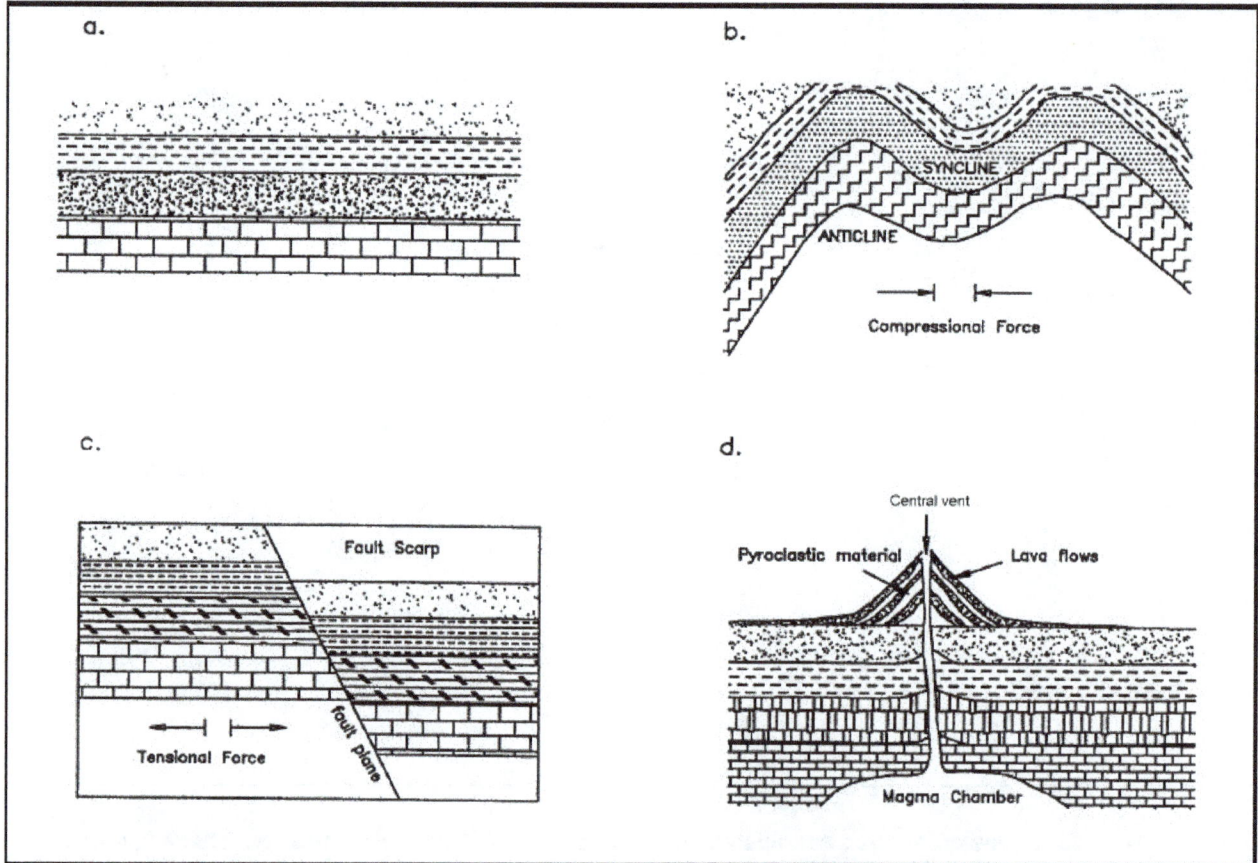

**Figure IV-1-7. Examples of tectonically produced features: (a) stable undeformed block; (b) symmetrical folding resulting from compressional forces; (c) normal faulting resulting from tensional forces; (d) composed of alternating layers of pyroclastic material (ash) and lava flows**

understand complex water motion in the nearshore areas of large bodies of water. Consequently, an understanding of the mechanics of wave motion is essential in the planning and design of coastal works.

(b)   A detailed discussion of water wave mechanics is presented in Parts II-1, II-2, and II-3. Bascom's (1964) *Waves and Beaches* is a general introduction to the subject for the nonspecialist.

(2) Tides.

(a)   The most familiar sea level changes are those produced by astronomical tides. *Tides* are a periodic rise and fall of water level caused by the gravitational interaction among the earth, moon, and sun. Because the earth is not covered by a uniform body of water, tidal ranges and periods vary from place to place and are dependent upon the natural period of oscillation for each water basin (Komar 1998). Tidal periods are characterized as diurnal (one high and one low per day), semidiurnal (two highs and two lows per day), and mixed (two highs and two lows with unequal heights) (Figure II-5-16). In the coastal zone, variations in topography, depth, seafloor sediment type, and lateral boundaries also affect the tide. For more background information and theory, refer to Part II-4.

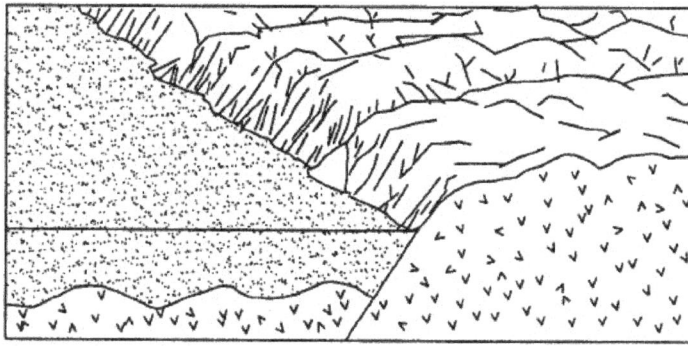

**Figure IV-1-8. Example of a fault coast exhibiting a prominent fault scarp**

**Figure IV-1-9. Example of a volcanic coast with numerous circular islands**

(b) The importance of tides to coastal geological processes is threefold. First, the periodic change in water level results in different parts of the foreshore being exposed to wave energy throughout the day. In regions with large tidal ranges, the water may rise and fall 10 m, and the shoreline may move laterally several kilometers between high and low water. This phenomenon is very important biologically because the ecology of tidal flats depends on their being alternately flooded and exposed. The geological significance is that various parts of the intertidal zone are exposed to erosion and deposition.

(c) Second, tidal currents themselves can erode and transport sediment. Generally, tidal currents become stronger near the coast and play an increasingly important role in local circulation (Knauss 1978). Because of the rotating nature of the tidal wave in many locations (especially inland seas and enclosed basins), ebb and flood currents follow different paths. As a result, residual motions can be highly important in terms of transport and sedimentation (Carter 1988). In inlets and estuaries, spatially asymmetric patterns of ebb and flood may cause mass transport of both water and sediment.

(d) Third, tides cause the draining and filling of tidal bays. These bays are found even in low-tide coasts such as the Gulf of Mexico. This process is important because it is related to the cutting and migration of tidal inlets and the formation of flood- and ebb-tidal shoals in barrier coasts. The exchange of seawater in and out of tidal bays is essential to the life cycle of many marine species.

(3) Energy-based classification of shorelines.

(a) Davies (1964) applied an energy-based classification to coastal morphology by subdividing the world's shores according to tide range. Hayes (1979) expanded this classification, defining five tidal categories for coastlines:

- Microtidal, < 1 m.

- Low-mesotidal, 1-2 m.

- High-mesotidal, 2-3.5 m.

- Low-macrotidal, 3.5-5 m.

- Macrotidal, > 5 m.

The Hayes (1979) classification was based primarily on shores with low to moderate wave power and was intended to be applied to trailing edge, depositional coasts.

(b) In the attempt to incorporate wave energy as a significant factor modifying shoreline morphology, five shoreline categories were identified based on the relative influence of tide range versus mean wave height (Figure IV-1-10) (Nummedal and Fischer 1978; Hayes 1979; Davis and Hayes 1984):

- Tide-dominated (high).

- Tide-dominated (low).

- Mixed-energy (tide-dominated).

- Mixed energy (wave-dominated).

- Wave-dominated.

(c) The approximate limit of barrier island development is in the field labeled "mixed energy (tide-dominated)." Notice that these fields cover a range of tide and wave heights. It is the relative effects of these processes that are important, not the absolute values. Also, at the lower end of the energy scales, there is a delicate balance between the forces; where tide-dominated, wave-dominated, or mixed-energy morphologies may develop with very little difference in wave or tide parameters. By extension, tidal inlets have sometimes been classified using this nomenclature.

(d) Continuing research has shown, however, that earlier approaches to classifying the coast on the basis of tidal and wave characteristics have been oversimplified because many other factors can play critical roles in determining shoreline morphology and inlet characteristics (Davis and Hayes 1984; Nummedal and Fischer 1978). Among these factors are:

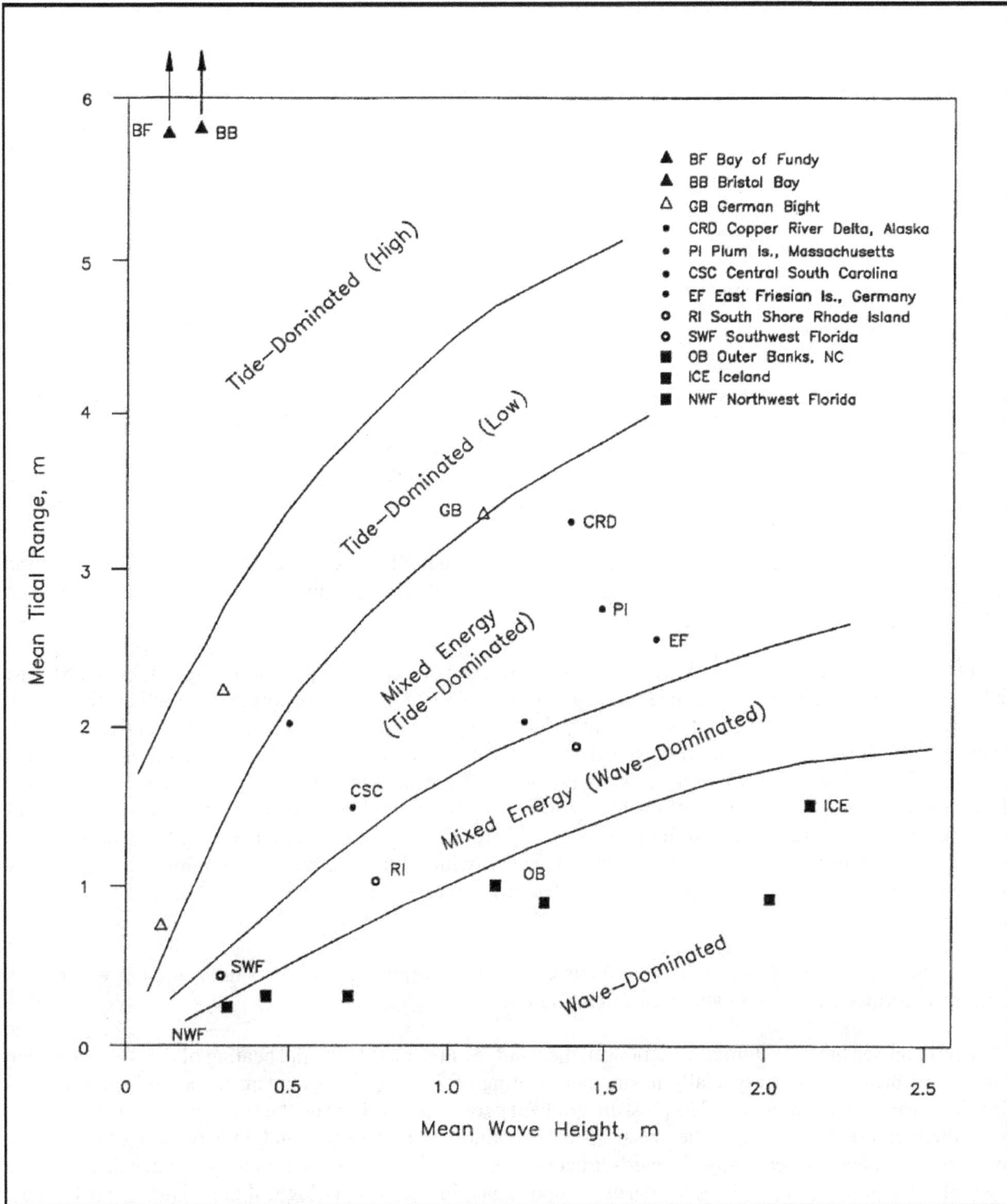

Figure IV-1-10. Energy-based classification of shorelines (from Hayes (1979))

● Physiographic setting and geology.

● Tidal prism.

● Sediment availability.

● Influence of riverine input.

● Bathymetry of the back-barrier bays.

● Meteorology and the influence of storm fronts.

(4) Meteorology. *Meteorology* is the study of the spatial and temporal behavior of atmospheric phenomena. *Climate* characterizes the long-term meteorologic conditions of an area, using averages and various statistics. Factors directly associated with climate such as wind, temperature, precipitation, evaporation, chemical weathering, and seawater properties all affect coastal geology. The shore is also affected by wave patterns that may be due to local winds or may have been generated by storms thousands of kilometers away. Fox and Davis (1976) is an introduction to weather patterns and coastal processes, and Hsu (1988) reviews coastal meteorology fundamentals.

(a) Wind. Wind is caused by pressure gradients, horizontal differences in pressure across an area. Wind patterns range in scale from global, which are generally persistent, to local and short duration, such as thunderstorms.

(b) Direct influence of wind. Wind has a great influence on coastline geomorphology, both directly and indirectly. The direct influence includes wind as an agent of erosion and transportation. It affects the coastal zone by eroding, transporting, and subsequently depositing sediment. Bagnold (1954) found that a proportional relationship exists between wind speed and rate of sand movement. The primary method of sediment transport by wind is through saltation, or the bouncing of sediment grains across a surface. Two coastal geomorphic features that are a direct result of wind are dunes and related blowouts (Pethick 1984). *Dunes* are depositional features whose form and size are a result of sediment type, underlying topography, wind direction, duration, and strength. *Blowouts* form when wind erodes an unvegetated area, thus removing the sand and leaving a depression or trough. These features are discussed in more detail in Parts III-4 and IV-2.

(c) Indirect effect. Wind indirectly affects coastal geomorphology as wind stress upon a water body causes the formation of waves and oceanic circulation.

(d) Land/sea breeze. Diurnal variations in the wind result from differential heating of the ocean and land surfaces. During the day, especially in summer, heating of the land causes the air to expand and rise, thus forming an area of low pressure. The pressure gradient between the water and the land surfaces causes a landward-directed breeze. At night, the ocean cools less rapidly than does the land, thus resulting in air rising over the ocean and subsequently seaward-directed breezes. These breezes are rarely greater than 8 m/sec (18 mph) and therefore do not have a great effect upon coastline geomorphology, although there may be some offshore-onshore transport of sediment on beaches (Komar 1998).

(e) Water level setup and setdown. Onshore winds cause a landward movement of the surface layers of the water and thus a seaward movement of deeper waters. Strong onshore winds, if sustained, may also cause increased water levels or setup. The opposite occurs during offshore winds.

(f) Seiches. *Seiches* are phenomena of standing oscillation that occur in large lakes, estuaries, and small seas in response to sudden changes in barometric pressure, violent storms, and tides. This condition causes the water within the basin to oscillate much like water sloshing in a bowl.

(5) Tropical storms. A *cyclone* is a system of winds that rotates about a center of low atmospheric pressure clockwise in the Southern Hemisphere and anti-clockwise in the Northern Hemisphere (Gove 1986). *Tropical storm* is a general term for a low-pressure, synoptic-scale[1] cyclone that originates in a tropical area. At maturity, tropical cyclones are the most intense and feared storms in the world; winds exceeding 90 m/sec (175 knots or 200 mph) have been measured, accompanied by torrential rain (Huschke 1959). By convention, once winds exceed 33 m/sec (74 mph), tropical storms are known as *hurricanes* in the Atlantic and eastern Pacific, *typhoons* in the western Pacific (Philippines and China Sea), and *cyclones* in the Indian Ocean.

(a) Tropical storms can cause severe beach erosion and destruction of shore-front property because elevated sea level, high wind, and depressed atmospheric pressure can extend over hundreds of kilometers. Tropical storms can produce awesome property damage (Table IV-1-3) and move vast quantities of sediment.

- The great Gulf of Mexico hurricane of 1900 inundated Galveston Island, killing over 6,000 residents (NOAA 1977-estimates range from 8-12,000 dead).

- During the September 1919 hurricane that struck the Florida Keys, 300 lives were lost in Key West, where winds were reported at 50 m/sec (110 mph). The final death toll of over 600, mostly in ships at sea, made this the third deadliest U.S. hurricane on record.

- The Great New England Hurricane that devastated Long Island and southern New England in September of 1938 killed 600 people and eliminated beach-front communities along the southern Rhode Island shore (Minsinger 1988). Survivors reported 15-m (50-ft) breakers sweeping over the Rhode Island barriers (Allen 1976). Shinnecock Inlet was cut through the Long Island barrier (Morang 1999).

- Hurricane Hugo hit the U.S. mainland near Charleston, South Carolina, on September 21, 1989, causing over $4 billion in damage, eroding the barriers, and producing other geologic changes up to 180 km north and 50 km south of Charleston (Davidson, Dean, and Edge 1990; Finkl and Pilkey 1991).

- A cyclone on June 9, 1998, inundated low-lying coastal plains and salt pans in northwest India. Over 14,000 people vanished into the Arabian Sea.

Simpson and Riehl (1981) have examined the effects of hurricanes in the United States. This work and Neumann et al. (1987) list landfall probabilities for the U.S. coastline. Tropical storms from 1871 to 1986 are plotted in Neumann et al. (1987). Tannehill (1956) identified all known Western Hemisphere hurricanes before the 1950's. Worldwide representative tropical storm tracks are shown in Figure IV-1-11 and Atlantic tracks in Figure II-5-29.

(b) The Saffir-Simpson Scale has been used for over 20 years by the U.S. National Weather Service to compare the intensity of tropical cyclones (Table IV-1-4). Cyclones are ranked into five categories based on maximum wind speed and the amount of damage that they cause (Figure IV-1-12), with 5 representing

---

[1] Synoptic-scale refers to large-scale weather systems as distinguished from local patterns such as thunderstorms.

Table IV-1-3
Damage Estimates and Payouts by Insurance Companies for U.S. Catastrophes

| Date | Event (Region of Greatest Influence) | Category | Insured loss (millions)[1] | Total Damage estimate (millions)[2] | Total Damage in 1996 U.S. Dollars (millions) |
|---|---|---|---|---|---|
| Aug. 1992 | Hurricane Andrew (Florida, Louisiana)[2] | 4 | $16,500 | 26,500 | 30,475 |
| Jan. 1994 | Northridge, Cailifornia, earthquake | | 12,500 | | |
| Sep. 1989 | Hurricane Hugo (South Carolina) | 4 | 4,195 | 7,000 | 8,490 |
| Sep. 1998 | Hurricane Georges | | 2.900 | ≈ 4,000 | |
| Oct. 1995 | Hurricane Opal (Florida, Alabama) | 3 | 2,100 | 3,000 | 3,100 |
| March 1993 | "Storm of the Century" (24 eastern states) | | 1,750 | 6,000 | |
| Aug. 1969 | Hurricane Camille (Mississippi, Louisiana) | 5 | - | 1,400 | 6,100 |
| Oct. 1991 | Oakland, California, fire | | 1,700 | | |
| Sep. 1996 | Hurricane Fran | 3 | 1,600 | 3,200 | 3,200 |
| Sep. 1992 | Hurricane Iniki (Hawaiian Islands) | — | 1,600 | 1,800 | 2,070 |
| Oct. 1989 | Loma Prieta, California, earthquake | | 960 | | |
| Dec. 1983 | Winter storms, 41 states | | 880 | | |
| April-May 1992 | Los Angeles riots | | 775 | | |
| April 1992 | Wind, hail, tornadoes, floods (Texas, Oklahoma) | | 760 | | |
| Sep. 1979 | Hurricane Frederic (Mississippi, Alabama) | 3 | 753 | 2,300 | 4,330 |
| Sep 1964 | Hurricane Dora (southeast Florida) | 2 | 250 | | 1,340 |
| Sep 1960 | Hurricane Donna (south Florida) | 4 | 300 | 387 | 2,100 |
| Sep. 1938 | Great New England Hurricane (Long Island, Rhode Island, Connecticut, Massachusetts) | 3 | 400[3] | 600 | — |

Notes:
1. Total damage costs exceed insurance values because municipal structures I ke roads are not insured. (Source: *The New York Times,* December 28, 1993, citing insurance industry and State of Florida sources; *Daytona Beach News-Journal* web edition, 12 June 1998).
2. Andrew caused vast property damage in south central Florida, when sub-standard structures were torn apart by the hurricane's winds. This proves that hurricanes are not merely coastal hazards, although coastal residents usually are at greatest risk because of the danger from storm surges. (Source: The Deadliest, Costliest, and Most Intense United States Hurricanes of this Century, NOAA Technical Memorandum NWS TPC-1 (www.nhc.noaa.gov/pastcast.html, 23 Dec 1998).
3. Multiplying the 1938 damage value by 10 gives a crude estimate in 1990's Dollars (data source: Minsinger 1988).

storms that cause catastrophic damage to structures. Only two Category 5 storms have hit the United States since record-keeping began: the 1935 Labor Day hurricane that hit the Florida Keys and killed 600 people, and Hurricane Camille, which devastated the Mississippi coast in 1969, killing 256 and causing 1.4 billion in damage (over $6 billion when converted to 1996 values).

(c) During tropical storms and other weather disturbances, water level changes are caused by two factors:

- *Barometric pressure.* Barometric pressure has an inverse relationship to sea level. As atmospheric pressure increases, the sea surface is depressed so that the net pressure on the seafloor remains constant. Inversely, as atmospheric pressure decreases, surface water rises. The magnitude of the "inverse barometer effect" is about 0.01 m for every millibar of difference in pressure, and in areas affected by tropical storms or hurricanes, the potential barometric surge may be as high as 1.5 m (Carter 1988).

Coastal Terminology and Geologic Environments

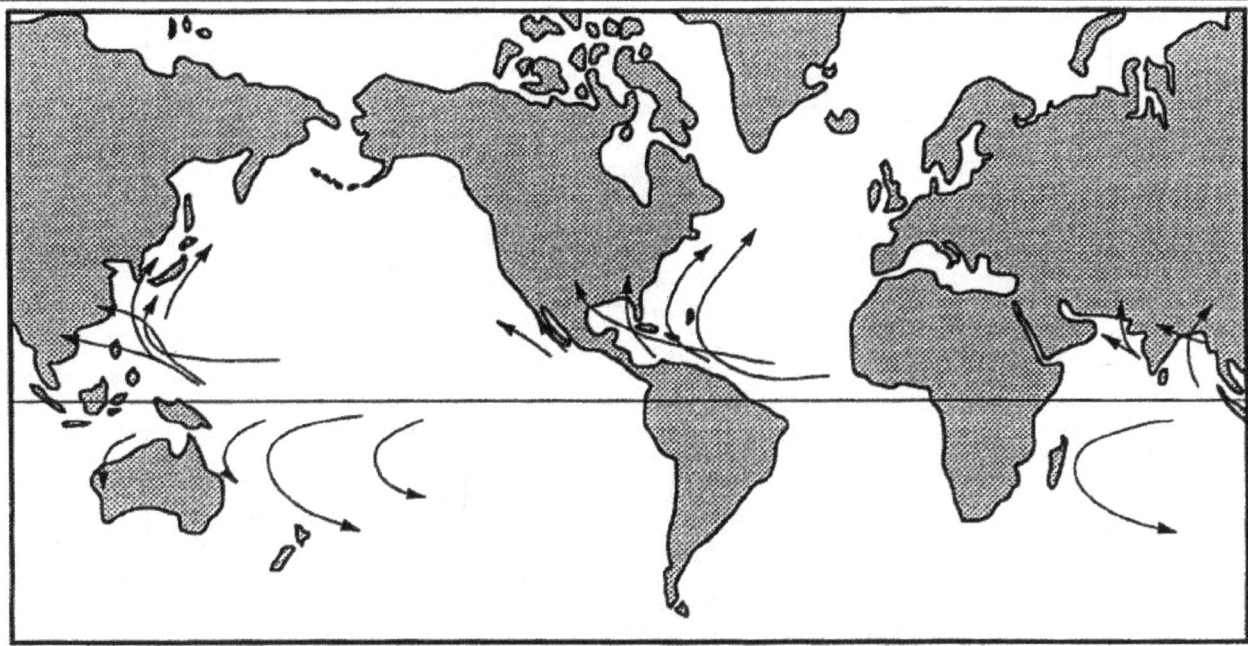

Figure IV-1-11.  Worldwide tropical storm pathways (from Cole (1980))

Table IV-1-4
Saffir-Simpson Damage-Potential Scale

| Scale Number (category) | Central pressure (millibars) | Wind speed (miles/hr) | Wind speed (m/sec) | Surge (ft) | Surge (m) | Damage |
|---|---|---|---|---|---|---|
| 1 | ≥980 | 74-95 | 33-42 | 4-5 | ~1.5 | Minimal |
| 2 | 965-979 | 96-110 | 43-49 | 6-8 | ~2-2.5 | Moderate |
| 3 | 945-964 | 111-130 | 50-58 | 9-12 | ~2.6-3.9 | Extensive |
| 4 | 920-944 | 131-155 | 59-69 | 13-18 | ~4-5.5 | Extreme |
| 5 | <920 | >155 | >69 | >18 | >5.5 | Catastrophic |

(From Hsu (1988); originally from Simpson and Riehl (1981))

- *Storm surges.*  In shallow water, winds can pile up water against the shore or drive it offshore.  Storm surges, caused by a combination of low barometric pressure and high onshore winds, can raise sea level several meters, flooding coastal property (Figures IV-1-13 and IV-1-14).  The Federal Emergency Management Agency (FEMA) determines base flood elevations for the coastal counties of the United States.  These elevations include still-water-level flood surges that have a 100-year return interval.  In light of rising sea level along most of the United States, it seems prudent that Flood Insurance Rate Maps be periodically adjusted (National Research Council 1987).  Besides wind forcing, ocean waves generated by storms can temporarily increase water levels tens of centimeters.  Analysis procedures for predicting surge heights are detailed in EM 1110-2-1412.

(6) Extratropical storms.  *Extratropical cyclones* (ET's) are cyclones associated with migratory fronts occurring in the middle and high latitudes (Hsu 1988).  Although hurricanes are the most destructive storms to pass over the U.S. Atlantic coast, less powerful ET's, more commonly known as winter storms or

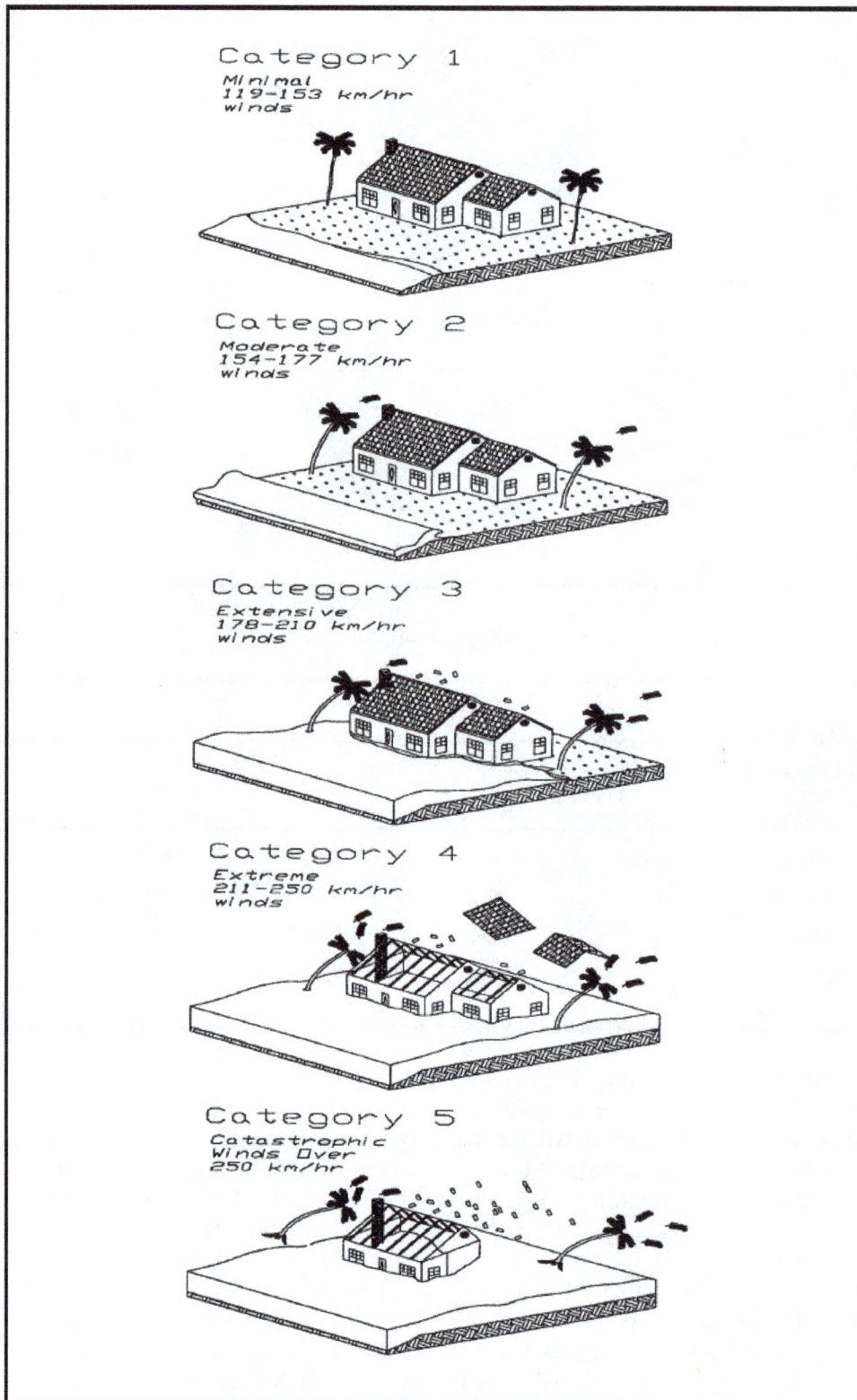

**Category 1**
Minimal
119-153 km/hr
winds

**Category 2**
Moderate
154-177 km/hr
winds

**Category 3**
Extensive
178-210 km/hr
winds

**Category 4**
Extreme
211-250 km/hr
winds

**Category 5**
Catastrophic
Winds Over
250 km/hr

Figure IV-1-12.  Graphical representation of the **Saffir-Simpson Scale**, showing the amount of damage that can be expected during different category hurricanes. Hurricane Andrew, in August 1992, was a **Category 5** storm that caused immense damage inland, without a storm surge (Figure modified from Associated Press figure printed in *Vicksburg Post,* 3 Aug 1995)

Figure IV-1-13.   Navarre Beach, Florida, November 1995.  The house in Santa Rosa Sound was lifted off its foundations and moved back hundreds of meters.  Many houses here were built on piles, but during Hurricane Opal, some piles were undermined, while at some properties, waves simply lifted buildings up off their supports.  The low area in the foreground is a washover channel

Figure IV-1-14.   Hurricane Opal damage at Navarre Beach, Florida, November 1995.  The sand underneath the concrete slab washed away, and the unsupported floors collapsed

"northeasters," have also damaged ships, eroded beaches, and taken lives. Northeasters are not as clearly defined as hurricanes and their wind speeds seldom approach hurricane strength. On the other hand, ET's usually cover broader areas than hurricanes and move more slowly; therefore, ET's can generate wave heights that exceed those produced by tropical storms (Dolan and Davis 1992).

(a) Most Atlantic northeasters occur from December through April. Dolan and Davis (1992) have tabulated historic ET's and calculated that the most severe ones are likely to strike the northeast coast in October and January.

(b) The *Halloween Storm* of October 1991 was one of the most destructive northeasters to ever strike the Atlantic coast. The system's lowest pressure dipped to 972 mb on October 30. Sustained winds about 40-60 knots persisted for 48 hr, generating immense seas and storm surges (Dolan and Davis 1992). Another famous northeaster was the *Ash Wednesday Storm* of 1962, which claimed 33 lives and caused great property damage.

(c) In early 1983, southern California was buffeted by the most severe storms in 100 years, which devastated coastal buildings and caused tremendous erosion. During one storm in January 1983, which coincided with a very high tide, the cliffs in San Diego County retreated as much as 5 m (Kuhn and Shepard 1984). Further north, the storm was more intense and cliff retreat of almost 30 m occurred in places. Kuhn and Shepard (1984) speculated that the unusual weather was linked to the eruption of El Chichon Volcano in the Yucatan Peninsula in March 1982. They noted that the 1983 storms in California were the most intense since the storms of 1884, which followed the August 27, 1883, explosion of Krakatoa.

(d) At this time, weather forecasters still have difficulty forecasting the development and severity of ET's. Coastal planners and engineers must anticipate that powerful storms may lash their project areas and need to apply conservative engineering and prudent development practices to limit death and property destruction.

c. *Biological factors.*

Coastal areas are normally the sites of intense biological activity. This is of enormous geological importance in some areas, while being insignificant and short-lived in others. Biological activity can be constructive; e.g., the growth of massive coral reefs, or it can be destructive, as when boring organisms help undermine sea cliffs. Remains from marine organisms having hard skeletal parts, usually composed of calcium carbonate, contribute to the sediment supply almost everywhere in the coastal environment. These skeletal contributions can be locally important and may even be the dominant source of sediment. Vegetation, such as mangroves and various grasses, plays an important role in trapping and stabilizing sediments. Growth of aquatic plants in wetlands and estuaries is critical in trapping fine-grained sediments, eventually leading to infilling of these basins (if balances between sediment supply and sea level changes remain steady). Kelp, particularly the larger species, can be an important agent of erosion and transportation of coarse detritus such as gravel and cobble. Biological coasts are discussed in greater detail in Part IV-2. Deltaic and estuarine processes, which are greatly influenced by biology, are discussed in Part IV-3.

## IV-1-6. Sea Level Changes

*a. Background.*

(1) General.

(a) Changes in sea level can have profound influence on the geology, natural ecology, and human habitation of coastal areas. A long-term and progressive rise in sea level has been cited as a major cause of erosion and property damage along our coastlines. Predicting and understanding this rise can guide coastal planners in developing rational plans for coastal development and the design, construction, and operation of structures and waterways.

(b) Many geomorphic features on contemporary coasts are the byproducts of the eustatic rise in sea level caused by Holocene climatic warming and melting of continental glaciers. Sea level has fluctuated throughout geologic time as the volume of ocean water has fluctuated, the shape of the ocean basins has changed, and continental masses have broken apart and re-formed.

(c) Sea level changes are the subject of active research in the scientific community and the petroleum industry. The poor worldwide distribution of tide gauges has hampered the study of recent changes (covering the past century) as most gauges were (and still are) distributed along the coasts of industrial nations in the Northern Hemisphere. Readers interested in this fascinating subject are referred to Emery and Aubrey's (1991) excellent book, *Sea Levels, Land Levels, and Tide Gauges.* This volume and Gorman (1991) contain extensive bibliographies. Tabular data and analyses of United States tide stations are printed in Lyles, Hickman, and Debaugh (1988), and worldwide Holocene sea level changes are documented in Pirazzoli (1991). Papers on sea level fluctuations and their effects on coastal evolution are presented in Nummedal, Pilkey, and Howard (1987). Engineering implications are reviewed in National Research Council (1987). Atmospheric carbon dioxide, climate change, and sea level are explored in National Research Council (1983). Houston (1993) discusses the state of uncertainty surrounding predictions on sea level change.

(2) Definitions. Because of the complexity of this topic, it is necessary to introduce the concepts of relative and eustatic sea level:

(a) *Eustatic* sea level change is caused by change in the relative volumes of the world's ocean basins and the total amount of ocean water (Sahagian and Holland 1991). It can be measured by recording the movement in sea surface elevation compared with some universally adopted reference frame. This is a challenging problem because eustatic measurements must be obtained from the use of a reference frame that is sensitive *only* to ocean water and ocean basin volumes. For example, highly tectonic areas (west coasts of North and South America; northern Mediterranean countries) are not suitable for eustatic sea level research because of frequent vertical earth movements (Mariolakos 1990). Tide gauge records from "stable" regions throughout the world have generated estimates of the recent eustatic rise ranging from 15 cm/century (Hicks 1978) to 23 cm/century (Barnett 1984).

(b) A *relative* change in water level is, by definition, a change in the elevation of the sea surface compared with some local land surface. The land, the sea, or both may have moved in *absolute* terms with respect to the earth's geoid. It is exceptionally difficult to detect absolute sea level changes because tide stations are located on land masses that have themselves moved vertically. For example, if both land and sea are rising at the same rate, a gauge will show that relative sea level (rsl) has not changed. Other clues, such as beach ridges or exposed beach terraces, also merely reflect their movement relative to the sea.

(3) Overview of causes of sea level change.

(a) Short-term sea level changes are caused by seasonal and other periodic or semi-periodic oceanographic factors. These include astronomical tides, movements of ocean currents, runoff, melting ice, and regional atmospheric variations. Included in this category are abrupt land level changes that result from volcanic activity or earthquakes. *Short-term* is defined here as an interval during which we can directly see or measure the normal level of the ocean rising or falling (a generation or 25 years). These factors are of particular pertinence to coastal managers and engineers, who are typically concerned with projects expected to last a few decades and who need to anticipate sea level fluctuations in their planning.

(b) Slow, secular sea and land level changes, covering time spans of thousands or millions of years, have been caused by glacioeustatic, tectonic, sedimentologic, climatologic, and oceanographic factors. Sea level was about 100 to 130 m lower during the last glacial epoch (Figure IV-1-15), about 15,000 years before present. Ancient shorelines and deltas can be found at such depths along the edge of the continental shelf (Suter and Berryhill 1985). Changes of this magnitude have been recorded during other geological epochs (Payton 1977).

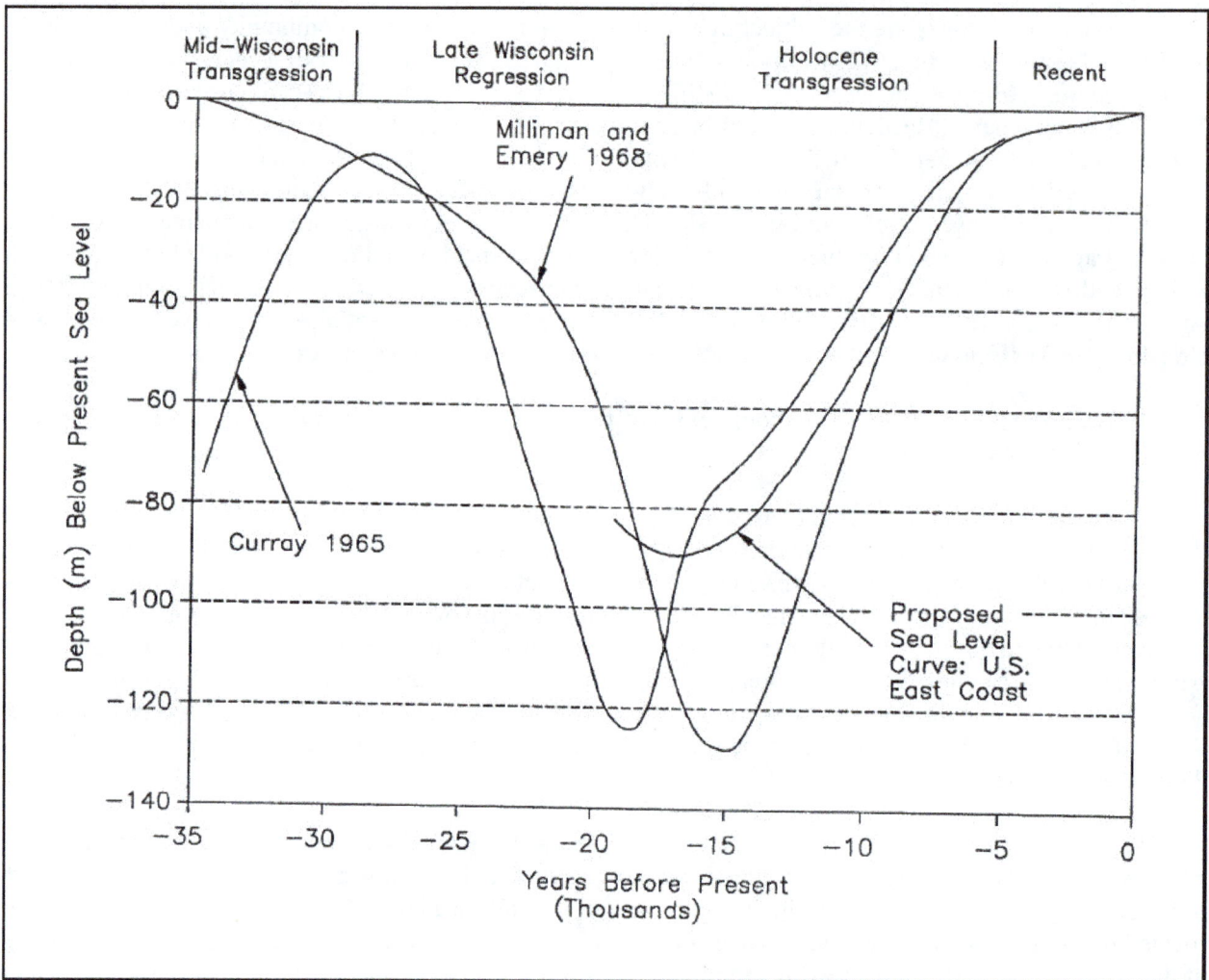

**Figure IV-1-15. Sea level fluctuations during the Pleistocene and Holocene epochs (adapted from Nummedal (1983); data from Dillon and Oldale (1978))**

(c) Table IV-1-5 lists long-term and short-term factors along with estimates of their effect on sea level. The following paragraphs discuss some factors in greater detail.

*b. Short-term causes of sea level change.*

(1) Seasonal sea level changes.

(a) The most common of the short-term variations is the seasonal cycle, which in most areas accounts for water level changes of 10 to 30 cm (and in some unusual cases - the Bay of Bengal - as much as 100 cm) (Komar and Enfield 1987). Seasonal effects are most noticeable near river mouths and estuaries. Variations in seasonal river flow may account for up to 21 percent of annual sea level variations in coastal waters (Meade and Emery 1971). Compared to the eustatic rise of sea level, estimated to be up to 20 cm/century, the seasonal factor may be a more important cause of coastal erosion because of its greater year-to-year influence (Komar and Enfield 1987).

(b) Over most of the world, lowest sea level occurs in spring and highest in autumn. Separating the individual factors causing the annual cycle is difficult because most of the driving mechanisms are coherent - occurring in phase with one another. Variations in atmospheric pressure drive most of the annual sea level change (Komar and Enfield 1987).

(2) West coast of North America.

(a) The west coast is subject to extreme and complicated water level variations. Short-term fluctuations are related to oceanographic conditions like the El Niño-Southern Oscillation. This phenomenon occurs periodically when equatorial trade winds in the southern Pacific diminish, causing a seiching effect that travels eastward as a wave of warm water. This raises water levels all along the U.S. west coast. Normally, the effect is only a few centimeters, but during the 1982-83 event, sea level rose 35 cm at Newport, Oregon (Komar 1992). Although these factors do not necessarily cause permanent geologic changes, engineers and coastal planners must consider their potential effects. The most recent El Niño event, during the winter of 1997-98, has been blamed for causing unusual weather in the western United States, including greatly increased rainfall in California and a warm winter in Oregon and Washington. Coastal geological changes caused by the El Niño are difficult to document. It has been argued (especially in the media) that increased rainfall in California caused more mudflows and bluff collapse than normal.

(b) Seasonal winter storms along the Pacific Northwest can combine with astronomical tides to produce elevated water levels over 3.6 m. During the severe storms of 1983, water levels were 60 cm over the predicted level.

(3) Rapid land level changes. Earthquakes are shock waves caused by abrupt movements of blocks of the earth's crust. A notable example occurred during the Great Alaskan Earthquake of 1964, when changes in shoreline elevations ranged from a 10-m uplift to a 2-m downdrop (Hicks 1972; Plafker and Kachadoorian 1966).

(4) Ocean temperature. Changes in the water temperature of upper ocean layers cause changes in water density and volume. As surface water cools, the density of seawater increases, causing a decrease in volume, thus lowering sea level. When temperature increases, the opposite reaction occurs. Variations in water temperature are not simply due to seasonal changes in solar radiation but are primarily caused by changes in offshore wind and current patterns.

Table IV-1-5
Sea Level Changes Along the Coastal Zone

| Short-Term (Periodic) Causes | Time scale (P = period) | Vertical Effect[1] |
|---|---|---|
| **Periodic Sea Level Changes** | | |
| Astronomical tides | 6-12 hr P | 0.2-10+ m |
| Long-period tides | | |
| Rotational variations (Chandler effect) | 14 month P | |
| **Meteorological and Oceanographic Fluctuations** | | |
| Atmospheric pressure | | |
| Winds (storm surges) | 1-5 days | Up to 5 m |
| Evaporation and precipitation | Days to weeks | |
| Ocean surface topography (changes in water density and currents) | Days to weeks | Up to 1 m |
| El Niño/southern oscillation | 6 mo every 5-10 yr | Up to 60 cm |
| **Seasonal Variations** | | |
| Seasonal water balance among oceans (Atlantic, Pacific, Indian) | | |
| Seasonal variations in slope of water surface | | |
| River runoff/floods | 2 months | 1 m |
| Seasonal water density changes (temperature and salinity) | 6 months | 0.2 m |
| **Seiches** | Minutes-hours | Up to 2 m |
| **Earthquakes** | | |
| Tsunamis (generate catastrophic long-period waves) | Hours | Up to 10 m |
| Abrupt change in land level | Minutes | Up to 10 m |

| Long-Term Causes | Range of Effect E = Eustatic; L = Local | Vertical Effect[1] |
|---|---|---|
| **Change in Volume of Ocean Basins** | | |
| Plate tectonics and seafloor spreading (plate divergence/convergence) and change in seafloor elevation (mid-ocean volcanism) | E | 0.01 mm/yr |
| Marine sedimentation | E | < 0.01 mm/yr |
| **Change in Mass of Ocean Water** | | |
| Melting or accumulation of continental ice | E | 10 mm/yr |
| Release of water from earth's interior | E | |
| Release or accumulation of continental hydrologic reservoirs | E | |
| **Uplift or Subsidence of Earth's Surface (Isostasy)** | | |
| Thermal-isostasy (temperature/density changes in earth's interior) | L | |
| Glacio-isostasy (loading or unloading of ice) | L | 1 cm/yr |
| Hydro-isostasy (loading or unloading of water) | L | |
| Volcano-isostasy (magmatic extrusions) | L | |
| Sediment-isostasy (deposition and erosion of sediments) | L | < 4 mm/yr |
| **Tectonic Uplift/Subsidence** | | |
| Vertical and horizontal motions of crust (in response to fault motions) | L | 1-3 mm/yr |
| **Sediment Compaction** | | |
| Sediment compression into denser matrix | L | |
| Loss of interstitial fluids (withdrawal of groundwater or oil) | L | ≤ 55 mm/yr[1] |
| Earthquake-induced vibration | L | |
| **Departure from Geoid** | | |
| Shifts in hydrosphere, aesthenosphere, core-mantle interface | L | |
| Shifts in earth's rotation, axis of spin, and precession of equinox | E | |
| External gravitational changes | E | |

[1]Effects on sea level are estimates only. Many processes interact or occur simultaneously, and it is not possible to isolate the precise contribution to sea level of each factor. Estimates are not available for some factors. (Sources: Emery and Aubrey (1991); Gornitz and Lebedeff (1987); Komar and Enfield (1987))

[1] Calculated using Shanghai as an example: 2.7 m subsidence between 1920 and 1970 (Baeteman 1994)

(5) Ocean currents. Because of changes in water density across currents, the ocean surface slopes at right angles to the direction of current flow. The result is an increase in height on the right side of the current (when viewed in the direction of flow) in the Northern Hemisphere and to the left in the Southern Hemisphere. The elevation change across the Gulf Stream, for example, exceeds 1 m (Emery and Aubrey 1991). In addition, major currents in coastal areas can produce upwelling, a process that causes deep colder water to move upward, replacing warmer surface waters. The colder upwelled water is denser, resulting in a regional decrease in sea level.

*c. Long-term causes of sea level change.*

(1) Tectonic instability. Regional, slow land level changes along the U.S. western continental margin affect relative long-term sea level changes. Parts of the coast are rising and falling at different rates. In Oregon, the northern coast is falling while the southern part is rising relative to concurrent relative sea level (Komar 1992).

(2) Isostacy. *Isostatic adjustment* is the process by which the crust of the Earth attains gravitational equilibrium with respect to superimposed forces (Emery and Aubrey 1991). If a gravitational imbalance occurs, the crust rises or sinks to correct the imbalance.

(a) The most widespread geologically rapid isostatic adjustment is the depression of land masses caused by glaciers and the rebounding caused by deglaciation. In Alaska and Scandinavia, contemporary uplift follows the depression of the crust caused by the Pleistocene ice sheets. Some areas of the Alaska coast (for example, Juneau) are rising over 1 cm/year, based on tide gauge records (Figure IV-1-16) (Lyles, Hickman, and Debaugh 1988).

(b) Isostatic adjustments have also occurred due to changes in sediment load on continental shelves and at deltas. The amount of sediment loading on shelves is not well determined but is probably about 4 mm/year. The effect is only likely to be important at deltas where the sedimentation rate is very high (Emery and Aubrey 1991).

(3) Sediment compaction.

(a) Compaction occurs when poorly packed sediments reorient into a more dense matrix. Compaction can occur because of vertical loading from other sediments, by draining of fluids from the sediment pore space (usually a man-made effect), by desiccation (drying), and by vibration.

(b) Groundwater and hydrocarbon withdrawal is probably the main cause of sediment compaction on a regional scale. Many of the world's great cities are located on coastal plains or on river mouth deltas. Because of the dense population and industrialization, vast quantities of groundwater have been pumped from the subsurface aquifers. The consequence is nearly instantaneous local land subsidence due to sediment compaction, transforming many of these great coastal cities into the sinking cities of the world (Baeteman 1994; see Table IV-1-6). Subsidence exceeding 8 m has been recorded in Long Beach, California, and over 6 m in the Houston-Freeport area (Emery and Aubrey 1991). In Galveston, the annual sea level rise shown on tide records is 0.6 cm/year (Figure IV-1-17) (Lyles, Hickman, and Debaugh 1988). Subsidence at Venice, Italy, caused by groundwater pumping, has been well-publicized because of the threat to architectural and art treasures. Fortunately, subsidence there appears to have been controlled now that alternate sources of water are being tapped for industrial and urban use (Emery and Aubrey 1991).

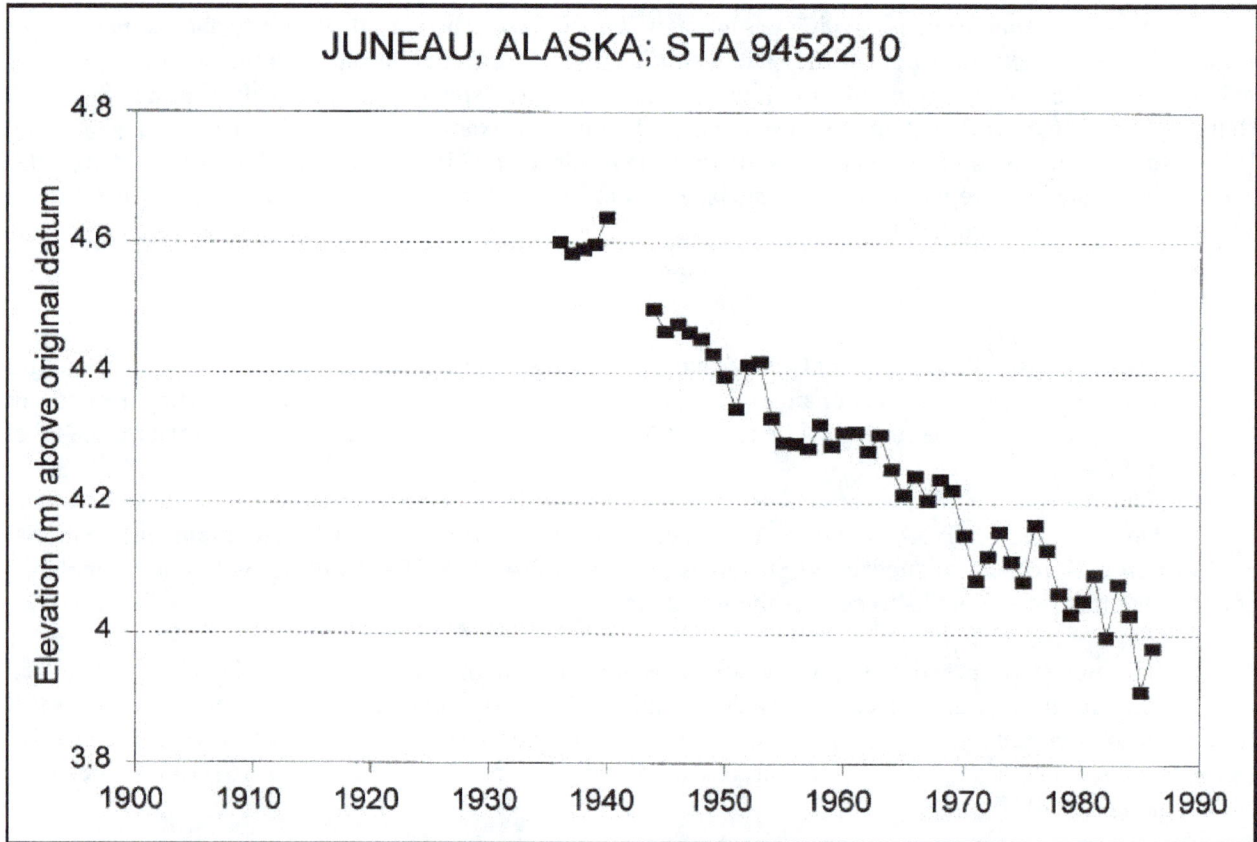

Figure IV-1-16. Yearly mean sea level changes at Juneau, Alaska, from 1936-1986. The fall in sea level shows the effects of isostatic rebound (data from Lyles, Hickman, and Debaugh (1988))

Table IV-1-6
Major World Cities with Recorded Subsidence[1]

| City or Region and Country | Subsidence (m) |
| --- | --- |
| Tokyo, Japan | 4.6 |
| Po Delta, Italy | 3.2 |
| Shanghai, China | 2.7 |
| Houston, USA | 2.7 |
| Tianjin, China | 2.5 |
| SW Taiwan | 2.4 |
| Taipei, Taiwan | 1.9 |
| Bangkok, Thailand | 1.6 |
| Ravenna, Italy | 1.2 |
| London, England | 0.35 |

[1] Records are not available for many other big cities (e.g., Jakarta, Hanoi, Haiphong, Rangoon, Manila).
From Baeteman (1994)

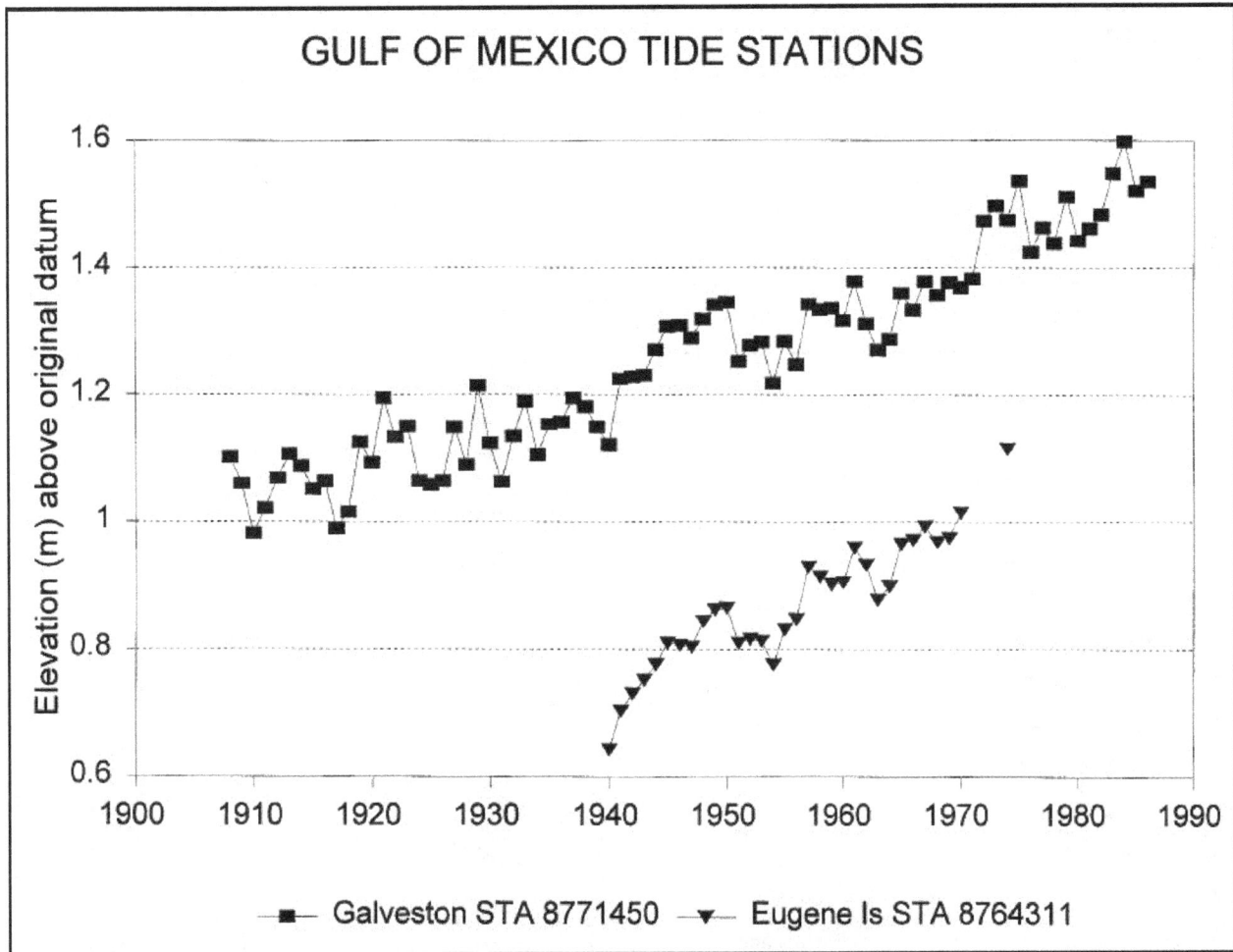

**Figure IV-1-17. Yearly mean sea level changes at Galveston, Texas, and Eugene Island, Louisiana. Subsidence of the land around Galveston may be caused by groundwater withdrawal and sediment compaction (data from Lyles, Hickman, and Debaugh (1988))**

(c) Significant subsidence occurs in and near deltas, where great volumes of fine-grained sediment accumulate rapidly. Land loss in the Mississippi delta has become a critical issue in recent years because of the loss of wetlands and rapid shoreline retreat. Along with natural compaction of underconsolidated deltaic muds and silts, groundwater and hydrocarbon withdrawal and river diversion might be factors contributing to the subsidence problems in southern Louisiana. Tide gauges at Eugene Island and Bayou Rigaud show that the rate of subsidence has increased since 1960 (Emery and Aubrey 1991). Change in rsl in the Mississippi Delta is about 15 mm/year, while the rate at New Orleans is almost 20 mm/year (data cited in Frihy (1992)).

*d. Geologic implications of sea level change.*

(1) Balance of sediment supply versus sea level change. Changes in sea level will have different effects on various portions of the world's coastlines, depending on conditions such as sediment type, sediment supply, coastal planform, and regional tectonics. The shoreline position in any one locale responds to the cumulative effects of the various sea level effects (outlined in Table IV-1-7). For simplicity, these factors can be subdivided into two broad categories: sediment supply and relative sea level (rsl) change. Ultimately, shoreline position is a balance between sediment availability and the rate that sea level changes

Table IV-1-7
Relative Effects of Sediment Supply Versus Sea Level Change on Shoreline Position[1]

| | | Relative Sea Level Change | | | | |
|---|---|---|---|---|---|---|
| | | Falling sea level | | Stable | Rising sea level | |
| | | Rapid | Slow | | Slow | Rapid |
| Sediment supply | Rapid net loss | Neutral | Slow retreat | Medium retreat | Rapid retreat[4] | Extra rapid retreat[2] |
| | Slow net loss | Slow advance | Neutral | Slow retreat | Medium retreat[6] | Rapid retreat |
| | Zero net change | Medium advance | Slow advance | Neutral[8] | Slow retreat[9] | Medium retreat |
| | Low net deposition | Rapid advance | Medium advance[10] | Slow advance[7] | Neutral[3,5] | Slow retreat |
| | Rapid net deposition | Extra rapid advance | Rapid advance[11] | Medium advance | Slow advance[1] | Neutral |

Examples of long-term (years) transgression or regression:

1. Mississippi River Delta - active distributary

2. Mississippi River Delta - abandoned distributary

3. Florida Panhandle between Pensacola and Panama City

4. Sargent Beach, TX

5. Field Research Facility, Duck, NC

6. New Jersey shore

7. Island of Hawaii - volcanic and coral sediment supply

8. Hawaiian Islands without presently active volcanoes

9. South shore of Long Island (sand trapped at inlets is balanced by man-made renourishment and bypassing)

10. Great Lakes during sustained fall in water levels

11. Alaska river mouths

[1] (Table based on a figure in Curray (1964))

(Table IV-1-7). For example, at an abandoned distributary of the Mississippi River delta, rsl is rising rapidly because of compaction of deltaic sediment. Simultaneously, wave action causes rapid erosion. The net result is extra rapid shoreline retreat (the upper right box in Table IV-1-7). The examples in the table are broad generalizations, and some sites may not fit the model because of unique local conditions.

(2) Historical trends. Historical records show the prevalence of shore recession around the United States during the past century (summarized by the National Research Council (1987):

(a) National average (unweighted) erosion rate: 0.4 m/year.

(b) Atlantic Coast: 0.8 m/year (with Virginia barrier islands exhibiting the highest erosion rates).

(c) Gulf Coast: 1.8 m/year (with highest erosion rate in Louisiana, 4.2 m/year).

(d) Pacific coastline: essentially stable (although more than half the shore is hard rock).

Bird (1976) claims that most sandy shorelines around the world have retreated during the past century. However, prograding shores are found where rivers supply excess sediment or where tectonic uplift is in progress.

(3) Specific coastal sites.

(a) Sandy (barrier) coasts. Several models predicting the effects of sea level rise on sandy coasts have been proposed. One commonly cited model is the Bruun rule. The Bruun rule and barrier migration models are discussed in Part IV-2.

(b) Cliff retreat. Cliff retreat is a significant problem in the Great Lakes, along the Pacific coast, and in parts of New England and New York. Increases in water level are likely to accelerate the erosion rate along Great Lakes shores (as shown by Hands (1983) for eastern Lake Michigan). However, along southern California, cliff retreat may be episodic, caused by unusually severe winter storms, groundwater and surface runoff, and, possibly, faulting and earthquakes, factors not particularly influenced by sea level (Kuhn and Shepard 1984). Crystalline cliffs are essentially stable because their response time is so much slower than that of sandy shores. Mechanisms of cliff erosion are discussed in Part IV-2.

(c) Marshes and wetlands. Marshes and mangrove forests fringe or back most of the Gulf and Atlantic coastlines. Marshes have the unique ability to grow upward in response to rising sea level. However, although marshes produce organic sediment, at high rates of rsl rise, additional sediment from outside sources is necessary to allow the marshes to keep pace with the rising sea. Salt marshes are described in Part IV-2-11. Paragraph IV-2-12 describes wetlands, coral and oyster reefs, and mangrove forest coasts. These shores have the natural ability to adjust to changing sea level as long as they are not damaged by man-made factors like urban runoff or major changes in sediment supply.

  e.  *Engineering and social implications of sea level change.*

(1) Eustatic sea level rise.

(a) Before engineering and management can be considered, a fundamental question must be asked: Is sea level still rising? During the last decade, the media has "discovered" global warming, and many politicians and members of the public are convinced that greenhouse gases are responsible for rising sea level and the increased frequency of flooding that occurs along the coast during storms. Most scientists accept the findings that the concentrations of greenhouse gasses in the atmosphere have increased greatly in the last century, largely due to industrial and automobile emissions. However, the link between increased gas in the atmosphere and changing sea level is much more difficult to model and verify. Wunsch (1996) has pointed out how difficult it is to separate myth from fact in the politically and emotionally charged issues of climate change and the oceans. The Environmental Protection Agency created a sensation in 1983 when it published a report linking atmospheric carbon dioxide to a predicted sea level rise of between 0.6 and 3.5 m (Hoffman, Keyes, and Titus 1983). Since then, predictions of the eustatic rise have been falling, and some recent evidence suggests that the rate may slow or even that eustatic sea level may drop in the future (Houston 1993).

(b) Possibly more reliable information on Holocene sea level changes can be derived from archaeological sites, wave-cut terraces, or organic material. For example, Stone and Morgan (1993) calculated an average rise of 2.4 mm/year from radiocarbon-dated peat samples from Santa Rosa Island, on the tectonically stable Florida Gulf coast. However, Tanner (1989) states that difficulties arise using all of these methods, and that calculated dates and rates may not be directly comparable.

(c) Based on an exhaustive study of tide records from around the world, Emery and Aubrey (1991) have concluded that it is not possible to assess if a *eustatic* rise is continuing because, while many gauges do record a recent rise in *relative* sea level, an equal number record a fall. Emery and Aubrey state (p. ix):

> In essence, we have concluded that 'noise' in the records produced by tectonic movements and both meteorological and oceanographic factors so obscures any signal of eustatic rise of sea level that the tide gauge records are more useful for learning about plate tectonics than about effects of the greenhouse heating of the atmosphere, glaciers, and ocean water.

They also state (p. 176):

> This conclusion should be no surprise to geologists, but it may be unexpected by those climatologists and laymen who have been biased too strongly by the public's perception of the greenhouse effect on the environment....Most coastal instability can be attributed to tectonism and documented human activities without invoking the spectre of greenhouse-warming climate or collapse of continental ice sheets.

(d) In summary, despite the research and attention devoted to the topic, the evidence about worldwide, eustatic sea level rise is inconclusive. Estimates of the rate of rise range from 0 to 3 mm/year, but some researchers maintain that it is not possible to discover a statistically reliable rate using tide gauge records. In late Holocene time, sea level history was much more complicated than has generally been supposed (Tanner 1989), suggesting that there are many perturbations superimposed on "average" sea level curves. Regardless, the topic is sure to remain highly controversial.

(2) Relative sea level (rsl) changes.

(a) The National Research Council's Committee on Engineering Implications of Changes in Relative Sea Level (National Research Council 1987) examined the evidence on sea level changes. They concluded that rsl, on statistical average, is rising at most tide gauge stations located on continental coasts around the world. In their executive summary, they concluded (p. 123):

> The risk of accelerated mean sea level rise is sufficiently established to warrant consideration in the planning and design of coastal facilities. Although there is substantial local variability and statistical uncertainty, average relative sea level over the past century appears to have risen about 30 cm relative to the East Coast of the United States and 11 cm along the West Coast, excluding Alaska, where glacial rebound has resulted in a lowering of relative sea level. Rates of relative sea level rise along the Gulf coast are highly variable, ranging from a high of more than 100 cm/century in parts of the Mississippi delta plain to a low of less than 20 cm/century along Florida's west coast.

However, they, too, noted the impact of management practices:

> Accelerated sea level rise would clearly contribute toward a tendency for exacerbated beach erosion. However, in some areas, anthropogenic effects, particularly in the form of poor sand management practices at channel entrances, constructed or modified for navigational purposes, have resulted in augmented erosion rates that are clearly much greater than would naturally occur. Thus, for some years into the future, sea level rise may play a secondary role in these areas.

(b) Figure IV-1-18 is a summary of estimates of local rsl changes along the U.S. coast (National Research Council 1987). Users of this map are cautioned that the values are based on tide records only from

Coastal Terminology and Geologic Environments

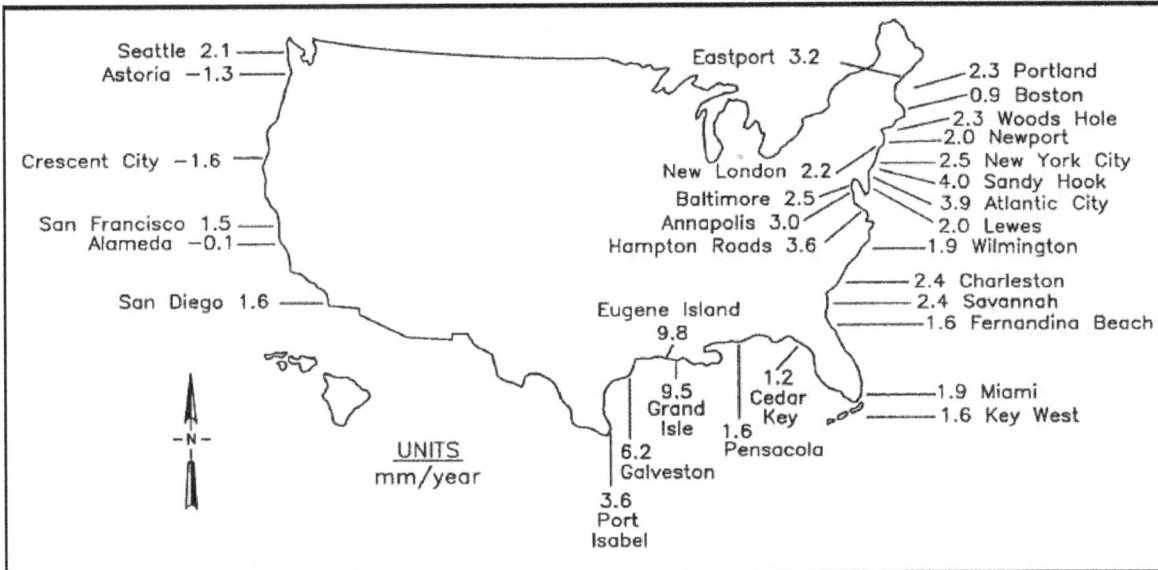

Figure IV-1-18. Summary of estimates of local rsl rise along the continental United States in millimeters per year. Values are based on tide gauge records during the period 1940-1980 (from National Research Council (1987))

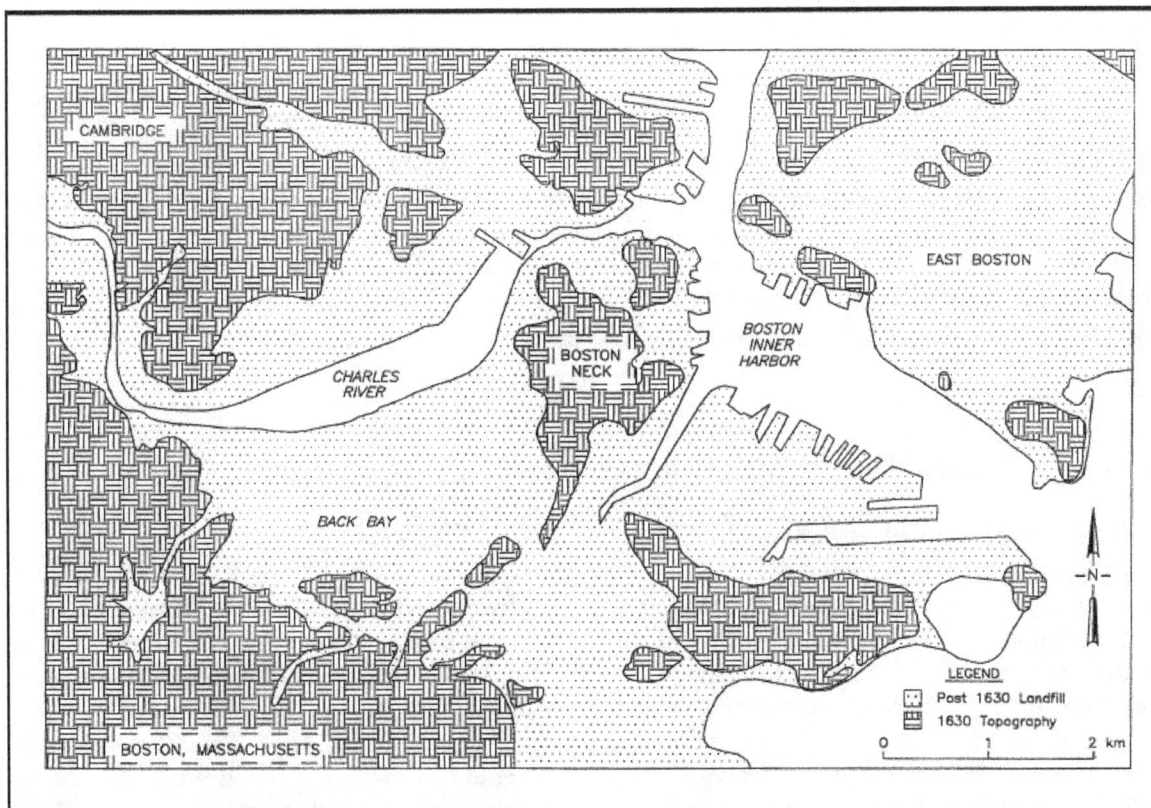

Figure IV-1-19. Landfilling in Boston, MA, since 1630 has more than doubled the urban area (unfortunately, at the expense of destroying what must have been highly productive wetlands) (from Rosen, Brenninkmeyer, and Maybury (1993))

1940-1980 and that much regional variability is evident. The figure provides general information only; for project use, detailed data should be consulted, such as the tide gauge statistics printed in Lyles, Hickman and Debaugh (1988) (examples of three tide stations are plotted in Figures IV-1-16 and IV-1-17) or the statistics available from the NOAA web site.

(3) Engineering response and policy.

(a) Whatever the academic arguments about eustatic sea level, engineers and planners must anticipate that changes in rsl may occur in their project areas and need to incorporate the anticipated changes in their designs and management plans.

(b) Because of the uncertainties surrounding sea level, the U.S. Army Corps of Engineers (USACE) has not endorsed a particular rise (or fall) scenario. Engineer Regulation (ER) 1105-2-100 (28 December 1990) states the official USACE policy on sea level rise. It directs that:

Feasibility studies should consider which designs are most appropriate for a range of possible future rates of rise. Strategies that would be appropriate for the entire range of uncertainty should receive preference over those that would be optimal for a particular rate of rise but unsuccessful for other possible outcomes.

Potential rsl rise should be considered in every coastal and estuarine (as far inland as the new head of tide) feasibility study that USACE undertakes. Project planning should consider what impact a higher rsl rise would have on designs based on local, historical rates.

(4) Impacts of rising sea level on human populations.

(a) Rising sea level raises the spectre of inundated cities, lost wetlands, and expensive reconstruction of waterways and ports. About 50 percent of the U.S. population lives in coastal counties (1980 census data reported in Emery and Aubrey (1991)), and the number is likely to increase. There has not been a long history of understanding and planning for sea level rise in the United States, but other countries, particularly Holland and China, have coped with the problem for thousands of years (National Research Council 1987). There are three principal ways that people could adapt to rising sea level:

● Retreat and abandonment.

● Armoring by erecting dikes and dams to keep out the sea.

● Construction on landfills and piers.

(b) Among the areas most susceptible to inundation caused by rise in rsl are deltas. Deltas are naturally sinking accumulations of sediment whose subaerial surface is a low-profile, marshy plain. Already, under present conditions, subsidence imposes especially severe hardships on the inhabitants in coastal Bangladesh (10 mm/year) and the Nile Delta (2 mm/year), two of the most densely populated regions on earth (Emery and Aubrey 1991). Even a slow rise in sea level could have devastating effects. How could these areas be protected? Thousands of kilometers of seawalls would be needed to protect a broad area like coastal Bangladesh from the sea and from freshwater rivers. Civil works projects on this scale seem unlikely in developing countries, suggesting that retreat will be the only recourse (National Research Council 1983). Nevertheless, despite the immense cost of large-scale coastal works, the Netherlands has reclaimed from the sea a large acreage of land, which is now used for towns and agriculture.

(c) Retreat can be either a gradual (planned or unplanned) process, or a catastrophic abandonment (National Research Council 1987). The latter has occurred in communities where buildings were not allowed to be rebuilt after they were destroyed or damaged by storms. The State of Texas followed this approach on Galveston Island after Hurricane Alicia in 1983 and the State of Rhode Island for some south shore communities after the Great New England Hurricane of 1938. Construction setback lines represent a form of controlled retreat. Seaward of setback lines, new construction is prohibited. City managers and coastal planners often have difficulty in deciding where setback lines should be located, and their decisions are bitterly contested by property developers who wish to build as close to the beach as possible.

(d) Most of the world's coastal cities are subject to inundation with even a modest rise of sea level. In 1990, of the 15 biggest "megacities" (population > 10 million, such as Tokyo, Shanghai, Buenos Aires, and Calcutta) 12 were in coastal areas (Young and Hale 1998). Unfortunately, 25 to 50 percent of these urban populations live in poverty, a situation that makes coastal management and planning for changing sea level very difficult. Nevertheless, irresistible political pressure will surely develop to defend cities against the rising sea because of the high concentration of valuable real estate and capital assets. Defense will most probably take the form of dikes like the ones that protect large portions of Holland and areas near Tokyo and Osaka, Japan, from flooding. Dikes would be needed to protect low-lying inland cities from rivers whose lower courses would rise at the same rate as the sea. Already, New Orleans (which is below sea level), Rotterdam, and other major cities located near river mouths are kept dry by levees. These levees might have to be raised under the scenario of rising sea level. Storm surge barriers, like the ones at New Bedford, Massachusetts, Providence, Rhode Island, and the Thames, below London, England, might have to be rebuilt to maintain an adequate factor of safety.

(e) Landfilling has historically been a common practice, and many coastal cities are partly built on landfill. Boston's waterfront, including the airport and the Back Bay, is built on 1800's fill (Figure IV-1-19; Whitehill 1968). Large areas around New York City, including parts of Manhattan and Brooklyn, have been filled since the 1600's (Leveson 1980). Venice, one of the world's great architectural treasures, occupies a cluster of low islands in the lagoon of Venice, at the head of the Adriatic Sea. In the early 1700's, Peter the Great built his monumental new capital of Saint Petersburg on pilings and fill in the estuary of the Neva River. Artificial land, which is usually low, is particularly susceptible to rising sea level. Although dikes and levees will probably be the most common means to protect cities threatened by the rising sea, there is a precedent in the United States for raising the level of the land surface where structures already exist: Seattle's downtown was raised about 3 m in the early 1900's to prevent tidal flooding. The elevated streets ran along the second floor of buildings, and the original sidewalks and store fronts remained one floor down at the bottom of open troughs. Eventually, the open sidewalks had to be covered or filled because too many pedestrians and horses were injured in falls.

*f. Changes in sea level - summary.*

(1) Changes in sea level are caused by numerous physical processes, including tectonic forces that affect land levels and seasonal oceanographic factors that influence water levels on various cycles (Table IV-1-5). Individual contributions of many of these factors are still unknown.

(2) Estimates of the eustatic rise in sea level range from 0 to 3 mm/year. Emery and Aubrey (1991) have strongly concluded that it is not possible to detect a statistically verifiable rate of eustatic sea level rise because of noise in the signals and because of the poor distribution of tide gauges worldwide.

(3) Arguments regarding eustatic sea level changes may be more academic than they are pertinent to specific projects. The rate of *relative* sea level change varies greatly around the United States. Coastal planners need to consult local tide gauge records to evaluate the potential movement of sea level in their project areas.

(4) In many areas, coastal management (mismanagement) practices have the greatest influence on erosion, and sea level changes are a secondary effect (Emery and Aubrey 1991; National Research Council 1987).

(5) USACE does not endorse a particular sea level rise (or fall) scenario. ER 1105-2-100 (28 December 1990) directs that feasibility studies must consider a range of possible future rates of sea level rise. Project planning should use local, historical rates of rsl change.

### IV-1-7. Cultural (Man-Made) Influences on Coastal Geology

*a. Introduction.* Man has modified many of the world's coastlines, either directly, by construction or dredging, or indirectly, as a result of environmental changes that influence sediment supply, runoff, or climate. Human activity has had the most profound effects on the coastal environment in the United States and the other industrial nations, but even shorelines in lesser-developed countries have not been immune to problems wrought by river diversion and loss of wetlands. The most common practices that significantly alter the coastal environment are the construction of coastal works such as jetties and groins and the development of property on and immediately inland of the beach. Historically, many cities have developed on the coast. Although originally most were located in bays or other protected anchorages, many have grown and spread to the open coast. Prominent United States examples include New York, Boston, San Diego, and Los Angeles. Still other communities originally began as resorts on barrier islands and have since grown into full-size cities; examples include Atlantic City, Ocean City, Virginia Beach, and Miami Beach. Land use practices well inland from the coast also often have important effects on coastal sedimentation. These factors are more difficult to detect and analyze because, sometimes, the affecting region is hundreds of kilometers inland. For example, dam construction can greatly reduce the natural supply of sediment brought to the coast by streams and rivers, while deforestation and agricultural runoff may lead to increased sediment load in rivers.

*b. Dams/Reservoirs.* In many coastal areas, the major source of sediment for the littoral system is from streams and rivers. Dams and reservoirs obstruct the transport of sediment to the littoral system by creating sediment traps. These structures also restrict peak flows, which reduce sediment transport of material that is available downstream of the structures. The net effect is sediment starvation of coastal areas that previously received riverine sediment. If the losses are not offset by new supplies, the results are shrinking beaches and coastal erosion (Schwartz 1982). The most prominent example is the accelerated erosion of the Nile Delta that has occurred since the Aswan Low Dam (1902) and the Aswan High Dam (1964) almost totally blocked the supply of sediment to the coast (Frihy 1992). The Rosetta promontory has been eroding at an average rate of 55 m/year between 1909 and the present. Loss of nutrient-laden silt from the Nile's annual spring floods has also had bad effects on agriculture in the Nile valley and delta and has damaged fisheries in the eastern Mediterranean. Portions of the southern California coast have also suffered this century from loss of fluvially supplied sediment (e.g., Point Arguello, cited by Bowen and Inman (1966)). Increased erosion of the Washington shore near Grays Harbor may be due to the loss of sediment from the Columbia River, which has been massively dammed since the 1930's and 1940's.

*c. Erosion control and coastal structures.* Coastal structures such as jetties, groins, seawalls, bulkheads, and revetments are probably the most dramatic cause of man-induced coastal erosion (*Shore Protection Manual* 1984). Any coastal structure will have some effect on local sediment dynamics, and in some cases, the effect may extend downdrift for many kilometers.

d. *Modification of natural protection.*

(1) Destructive effects. The destruction of dunes and beach vegetation, development of backshore areas, and construction on the back sides of barrier islands can increase the occurrence of overwash during storms. In many places, sand supply has diminished because much of the surface area of barriers has been paved and covered with buildings. The result has been backshore erosion and increased barrier island breaching. In most coastal areas of the United States, one need merely visit the local beaches to see examples of gross and callous coastal development where natural protection has been compromised. Carter (1988) reviews examples from the United Kingdom. Serious damage has occurred to biological shores around the world as a result of changes in runoff and sediment supply, increased pollution, and development.

(2) Constructive efforts. Sand dunes are often stabilized using vegetation and sand fences. Dunes afford protection against flooding of low-lying areas. Dunes are also stabilized to prevent sand from blowing over roads and farms. Dunes are discussed in Part IV-2-6.

e. *Beach renourishment (fill).* An alternative for restoring beaches without constructing groins or other hard structures is to bring sand to the site from offshore by dredges or from inland sources by truck. This is the only coastal management that actually adds sand back into the littoral system (Pope 1997). Although conceptually renourishment seems simple enough, in practice, the planning, design, application, and maintenance of beach renourishment projects are sophisticated engineering and geologic procedures. For design and monitoring information, the reader is referred to Part V-4, Tait (1993), and Stauble and Kraus (1993). *Shore and Beach,* Vol 61, No. 1 (January 1993) is a special issue devoted to the beach renourishment project at Ocean City, Maryland. Stauble et al. (1993) evaluate the Ocean City project in detail. Krumbein (1957) is a classic description of sediment analysis procedures for specifying beach fills. One of the most successful U.S. renourishment projects has been at Miami Beach, Florida (reviewed in Carter (1988)).

f. *Mining.*

(1) Beach mining can directly reduce the amount of sediment available to the littoral system. In most areas of the United States, beach sand can no longer be exploited for commercial purposes because sand is in short supply, and the health of dunes and biological communities depends vitally on the availability of sand. Strip mining can indirectly affect the coast due to increased erosion, which increases sediment carried to the sea by rivers (unless the sediment is trapped behind dams).

(2) In Britain, an unusual situation developed at Horden, County Durham, where colliery waste was dumped on the shore. The waste material formed a depositional bulge in the shore. As the sediment from Horden moved downcoast, it was sorted, with the less dense coal forming a surface placer on the beach that is commercially valuable (Carter 1988).

g. *Stream diversion.*

(1) Stream diversion, both natural and man-made, disrupts the natural sediment supply to areas that normally receive fluvial material. With diversion for agriculture or urban use, the results are similar to those produced by dams: sediment that normally would be carried to the coast remains trapped upriver. Its residence time in this artificial storage, decades or centuries, may be short on geological time scales but is long enough to leave a delta exposed to significant erosion.

(2) Natural diversion occurs when a river shifts to a new, shorter channel to the sea, abandoning its less efficient former channel. An example of this process is the gradual occupation of the Atchafalaya watershed by the Mississippi River. If this process were to continue to its natural conclusion, the present Balize ("Birdfoot") delta would be abandoned, causing it to erode at an ever faster rate, while a new delta

would form in Atchafalaya Bay (Coleman 1988). The evolution of the Mississippi River is discussed in Part IV-3-3.

*h. Agriculture.* Poor farming practices lead to exposure of farmlands and increased erosion rates. Eroded soil is easily carried away by streams and rivers and is ultimately deposited in estuaries and offshore. The consequence of this process is progradation of the depositional areas. If rivers have been dammed, the sediment load is trapped behind the dams in the artificial lakes, and in that case does not get carried to the open sea.

*i. Forestry.* Deforestation is a critical problem in many developing nations, where mountainsides, stripped of their protective trees, erode rapidly. The soil is carried to the sea, where local coastlines prograde temporarily, but upland areas are left bereft of invaluable topsoil, resulting in human poverty and misery and in the loss of animal habitat. Reckless slash-and-burn practices have destroyed many formerly valuable timber resources in Central America, and some southeast Asian countries have already cut down most of their trees (Pennant-Rea 1994). Fortunately, Malaysia and Indonesia are beginning to curb illegal timber cutting and export, a trend which hopefully will spread to other countries. Unfortunately, the financial turmoil that engulfed Asia in 1998 will probably set back efforts to promote responsible resource management.

## IV-1-8. References

**EM 1110-2-1412**
Storm Surge Analysis and Design Water Level Determinations

**Allen 1976**
Allen, E. S. 1976. *A Wind to Shake the World,* Little Brown & Co., New York, NY.

**Baeteman 1994**
Baeteman, C. 1994. "Subsidence in Coastal Lowlands Due to Groundwater Withdrawal: The Geological Approach; Coastal Hazards, Perception, Susceptibility and Mitigation," C. W. Finkl, Jr., ed., *Journal of Coastal Research* Special Issue No. 12, pp 61-75.

**Bagnold 1954**
Bagnold, R. A. 1954. *The Physics of Blown Sand and Desert Dunes,* 2nd ed., Methuen, London, UK.

**Barnett 1984**
Barnett, T. P. 1984. "The Estimation of 'Global' Sea Level: A Problem of Uniqueness," *Journal of Geophysical Research,* Vol 89, No C5, pp 7980-7988.

**Bascom 1964**
Bascom, W. 1964. *Waves and Beaches, the Dynamics of the Ocean Surface,* Doubleday & Co., Garden City, NY.

**Bates and Jackson 1984**
Bates, R. L., and Jackson, J. A. 1984. *Dictionary of Geologic Terms,* 3rd ed., Anchor Press/Doubleday, Garden City, NY.

**Bird 1976**
Bird, E. C. F. 1976. "Shoreline Changes During the Past Century," *Proceedings of the 23rd International Geographic Congress, Moscow,* Pergamon, Elmsford, NY.

**Bowen and Inman 1966**

Bowen, A. J., and Inman, D. L. 1966. "Budget of Littoral Sands in the Vicinity of Point Arguello, California," Technical Memorandum, Coastal Engineering Research Center, U.S. Army Engineer Waterways Experiment Station, Vicksburg, MS.

**Carter 1988**

Carter, R. W. G. 1988. *Coastal Environments: An Introduction to the Physical, Ecological, and Cultural Systems of Coastlines,* Academic Press, London, UK.

**Coastal States Organization 1997**

Coastal States Organization. 1997. "Putting the Public Trust Doctrine to Work - the Application of the Public Trust Doctrine to the Management of Lands, Water, and Living Resources of the Coastal States," Washington, DC.

**Cole 1980**

Cole, F. W. 1980. *Introduction to Meteorology,* John Wiley and Sons, Inc., New York, NY.

**Coleman 1988**

Coleman, J. M. 1988. "Dynamic Changes and Processes in the Mississippi River Delta," *Bulletin of the Geological Society of America,* Vol 100, pp 999-1015.

**Curray 1964**

Curray, J. R. 1964. "Transgressions and Regressions," *Papers in Marine Geology: Shepard Commemorative Volume*, R. L. Mills, ed., MacMillan, New York, NY.

**Curray 1965**

Curray, J. R. 1965. "Late Quaternary History, Continental Shelves of the United States," *The Quaternary of the United States,* H. E. Wright, Jr. and D. G. Frey, eds., Princeton University Press, Princeton, NJ, pp 713-735.

**Davidson, Dean, and Edge 1990**

Davidson, M. A., Dean, R. G., and Edge, B. L. 1990. *Shore and Beach,* Vol 58, No. 4 (Special issue dedicated to Hurricane Hugo papers).

**Davies 1964**

Davies, J. L. 1964. "A Morphogenic Approach to World Shorelines," *Zeitschrift für Geomorphology*, Vol 8, pp 27-42.

**Davis and Hayes 1984**

Davis, R. A., Jr., and Hayes, M. O. 1984. "What is a Wave-Dominated Coast?, Hydrodynamics and Sedimentation in Wave-Dominated Coastal Environments," B. Greenwood and R. A. Davis, Jr., eds., *Marine Geology*, Vol 60, pp 313-329.

**Dillon and Oldale 1978**

Dillon, W. D., and Oldale, R. N. 1978. "Late Quaternary Sea Level Curve: Reinterpretation Based on Glacio-Eustatic Influence," *Geology,* Vol 6, pp 56-60.

**Dolan and Davis 1992**

Dolan, R., and Davis, R. E. 1992. "Rating Northeasters," *Mariners Weather Log,* Vol 36, No 1, pp 4-11.

**Ellis 1978**
Ellis, M. Y. 1978. *Coastal Mapping Handbook.* Department of the Interior, U.S. Geological Survey and U.S. Department of Commerce, National Ocean Service and Office of Coastal Zone Management, U.S. Government Printing Office, Washington, DC.

**Emery and Aubrey 1991**
Emery, K. O., and Aubrey, D. G. 1991. *Sea Levels, Land Levels, and Tide Gauges,* Springer-Verlag, New York, NY.

**Finkl and Pilkey 1991**
Finkl, C. W., and Pilkey, O. H. 1991. "Impacts of Hurricane Hugo: September 10-22, 1989," *Journal of Coastal Research,* Special Issue No. 8.

**Flint 1971**
Flint, R. F. 1971. *Glacial and Quaternary Geology,* John Wiley and Sons, New York, NY.

**Fox and Davis 1976**
Fox, W. T., and Davis, R. A., Jr. 1976. "Weather Patterns and Coastal Processes," *Beach and Nearshore Sedimentation,* R. A. Davis, Jr., and R. L. Ethington, eds., Society of Economic Paleontologists and Mineralogists Special Publication No. 24, Tulsa, OK.

**Frihy 1992**
Frihy, O. E. 1992. "Sea-Level Rise and Shoreline Retreat of the Nile Delta Promontories, Egypt," *Natural Hazards,* Vol 5, pp 65-81.

**Gorman 1991**
Gorman, L. T. 1991. "Annotated Bibliography of Relative Sea Level Change," Technical Report CERC-91-16, U.S. Army Engineer Waterways Experiment Station, Vicksburg, MS.

**Gorman, Morang, and Larson 1998**
Gorman, L. T., Morang, A., and Larson, R. L. 1998. "Monitoring the Coastal Environment; Part IV: Mapping, Shoreline Change, and Bathymetric Analysis," *Journal of Coastal Research,* Vol 14, No. 1, pp 61-92.

**Gornitz and Lebedeff 1987**
Gornitz, V., and Lebedeff, S. 1987. "Global Sea-Level Changes During the Past Century," *Sea-Level Fluctuations and Coastal Evolution,* D. Nummedal, O. H. Pilkey, and J. D. Howard, eds., Special Publication No. 41, Society of Economic Paleontologists and Mineralogists, Tulsa, OK, pp 3-16.

**Gove 1986**
Gove, P. B., ed. 1986. *Webster's Third International Dictionary,* Merriam-Webster, Springfield, MA.

**Hands 1983**
Hands, E. B. 1983. "The Great Lakes as a Test Model for Profile Response to Sea Level Changes," Chapter 8 in *Handbook of Coastal Processes and Erosion,* P. D. Komar, ed., CRC Press, Inc., Boca Raton, FL. (Reprinted in Miscellaneous Paper CERC-84-14, Coastal Engineering Research Center, U.S. Army Engineer Waterways Experiment Station, Vicksburg, MS.)

**Hayes 1979**
Hayes, M. O. 1979. "Barrier Island Morphology as a Function of Tidal and Wave Regime," *Barrier Islands from the Gulf of St. Lawrence to the Gulf of Mexico*, S. P. Leatherman, ed., Academic Press, New York, NY, pp 1-29.

**Hicks 1972**
Hicks, S. D. 1972. "Changes in Tidal Characteristics and Tidal Datum Planes," *The Great Alaska Earthquake of 1964*, Oceanography and Coastal Engineering, National Academy of Sciences, Washington, DC, pp 310-314.

**Hicks 1978**
Hicks, S. D. 1978. "An Average Geopotential Sea Level Series for the United States," *Journal of Geophysical Research,* Vol 83, No. C3, pp 1377-1379.

**Hicks 1984**
Hicks, S. D. 1984. *Tide and Current Glossary.* NOAA/National Ocean Service, Rockville, MD.

**Hoffman, Keyes, and Titus 1983**
Hoffman, J. S., Keyes, D., and Titus, J. G. 1983. "Projecting Future Sea Level Rise; Methodology, Estimates to the Year 2100, and Research Needs," Report 230-09-007, U.S. Environmental Protection Agency, Washington, DC.

**Houston 1993**
Houston, J. R. 1993. "Responding to Uncertainties in Sea Level Rise," *The State of Art of Beach Nourishment, Proceedings of the 1993 National Conference on Beach Preservation Technology,* The Florida Shore & Beach Preservation Association, Tallahassee, FL, pp 358-372.

**Hsu 1988**
Hsu, S. A. 1988. *Coastal Meteorology*, Academic Press, Inc., San Diego, CA.

**Huschke 1959**
Huschke, R. E., ed. 1959. *Glossary of Meteorology,* American Meteorology Society, Boston, MA.

**International Hydrographic Bureau 1990**
International Hydrographic Bureau. 1990. *Hydrographic Dictionary, Part I.* International Hydrographic Bureau, Monaco.

**Knauss 1978**
Knauss, J. A. 1978. *Introduction to Physical Oceanography,* Prentice-Hall, Englewood Cliffs, NJ.

**Komar 1998**
Komar, P. D. 1976. *Beach Processes and Sedimentation,* Prentice-Hall, Englewood Cliffs, NJ.

**Komar 1992**
Komar, P. D. 1992. "Ocean Processes and Hazards Along the Oregon Coast," *Oregon Geology*, Vol 54, No. 1, pp 3-19.

**Komar and Enfield 1987**
Komar, P. D., and Enfield, D. B. 1987. "Short-Term Sea-Level Changes on Coastal Erosion," *Sea-Level Fluctuations and Coastal Evolution*, Special Publication No. 41, D. Nummedal, O. H. Pilkey, and J. D. Howard, eds., Society of Economic Paleontologists and Mineralogists, Tulsa, OK, pp 17-28.

**Kraft and Chrzastowski 1985**
Kraft, J. C., and Chrzastowski, M. J. 1985. "Coastal Stratigraphic Sequences," *Coastal Sedimentary Environments,* Davis, R. A., Jr., ed., Springer-Verlag, New York, NY, pp 625-663.

**Krumbein 1957**
Krumbein, W. C. 1957. "A Method for Specification of Sand for Beach Fills," Technical Memorandum No. 102, Beach Erosion Board, U.S. Army Engineer Waterways Experiment Station, Vicksburg, MS.

**Kuhn and Shepard 1984**
Kuhn, G. G., and Shepard, F. P. 1984. *Sea Cliffs, Beaches, and Coastal Valleys of San Diego County; Some Amazing Histories and Some Horrifying Implications,* University of California Press, Berkeley, CA.

**Larson, Morang, and Gorman 1997**
Larson, R. L., Morang, A., and Gorman, L. T. 1997. "Monitoring the Coastal Environment; Part II: Sediment Sampling and Geotechnical Methods," *Journal of Coastal Research,* Vol 13, No. 2, pp 308-330.

**Leveson 1980**
Leveson, D. 1980. *Geology and the Urban Environment,* Oxford University Press, New York, NY.

**Lyles, Hickman, and Debaugh 1988**
Lyles, S. D., Hickman, L. E., Jr., and Debaugh, H. A., Jr. 1988. "Sea Level Variations for the United States, 1855-1986," U.S. Department of Commerce, National Oceanic and Atmospheric Administration, National Ocean Service, Rockville, MD.

**Mariolakos 1990**
Mariolakos, I. 1990. "The Impact of Neotectonics with Regard to Canals, Pipelines, Dams, Open Reservoirs, etc. in Active Areas: The Case of the Hellenic Arc," *Greenhouse Effect, Sea Level and Drought, Proceedings of the NATO Advanced Research Workshop on Geohydrological Management of Sea Level and Mitigation of Drought (1989),* R. Paepe, R. W. Fairbridge, and S. Jelgersma, eds., Kluwer Academic Publishers, Dordrecht, The Netherlands, pp 427-438.

**Meade and Emery 1971**
Meade, R. H., and Emery, K. O. 1971. "Sea-Level as Affected by River Runoff, Eastern United States," *Science,* Vol 173, pp 425-428.

**Milliman and Emery 1968**
Milliman, J. D., and Emery, K. O. 1968. "Sea Levels During the Past 35,000 Years," *Science,* Vol 162, pp 1121-1123.

**Minsinger 1988**
Minsinger, W. E., ed. 1988. *The 1938 Hurricane, an Historical and Pictorial Summary,* Blue Hill Observatory, East Milton, MA.

**Morang, Larson, and Gorman 1997a**

Morang, A., Larson, R. L., and Gorman, L. T. 1997a. "Monitoring the Coastal Environment; Part I: Waves and Currents," *Journal of Coastal Research,* Vol 13, No. 1, pp 111-133.

**Morang, Larson, and Gorman 1997b**

Morang, A., Larson, R. L., and Gorman, L. T. 1997b. "Monitoring the Coastal Environment; Part III: Geophysical and Research Methods," *Journal of Coastal Research,* Vol 13, No. 4, pp 1964-1085.

**Morang 1999**

Morang, A. 1999. "Coastal Inlets Research Program, Shinnecock Inlet, New York, Site Investigation, Report 1, Morphology and Historical Behavior," Technical Report CHL-98-32, U.S. Army Engineer Waterways Experiment Station, Vicksburg, MS.

**Mossa, Meisburger, and Morang 1992**

Mossa, J., Meisburger, E. P., and Morang, A. 1992. "Geomorphic Variability in the Coastal Zone," Technical Report CERC-92-4, U.S. Army Engineer Waterways Experiment Station, Vicksburg, MS.

**National Research Council 1983**

National Research Council, Board on Atmospheric Sciences and Climate. 1983. *Changing Climate, Report of the Carbon Dioxide Assessment Committee,* National Academy Press, Washington, DC.

**National Research Council 1987**

National Research Council, Committee on Engineering Implications of Changes in Relative Mean Sea Level. 1987. *Responding to Changes in Sea Level,* National Academy Press, Washington, DC.

**Neumann, Jarvinen, Pike, and Elms 1987**

Neumann, C. J., Jarvinen, B. R., Pike, A. C., and Elms, J. D. 1987. *Tropical Cyclones of the North Atlantic Ocean, 1871-1986,* Third rev., Historical Climatology Series 6-2, National Climatic Data Center, Asheville, NC.

**NOAA 1977**

National Oceanic and Atmospheric Administration. 1977. "Some Devastating North Atlantic Hurricanes of the 20th Century," Booklet NOAA/PA 77019, U.S. Government Printing Office, Washington, DC.

**Nummedal 1983**

Nummedal, D. 1983. "Barrier Islands," *CRC Handbook of Coastal Processes and Erosion*, P. D. Komar, ed., CRC Press, Inc., Boca Raton, FL, pp 77-121.

**Nummedal and Fischer 1978**

Nummedal, D., and Fischer, I. A. 1978. "Process-Response Models for Depositional Shorelines: The German and the Georgia Bights," *Proceedings of the Sixteenth Conference on Coastal Engineering*, American Society of Civil Engineers, New York, NY, pp 1215-1231.

**Nummedal, Pilkey, and Howard, eds. 1987**

Nummedal, D., Pilkey, O. H., and Howard, J. D., eds. 1987. *Sea-Level Fluctuations and Coastal Evolution,* Special Publication No. 41, Society of Economic Paleontologists and Mineralogists, Tulsa, OK.

**Orme 1985**

Orme, A. R. 1985. "California," *The World's Coastline,* E. C. Bird, and M. L. Schwartz, eds., Van Nostrand Reinhold, New York, NY, pp 27-36.

**Payton 1977**
Payton, C. E., ed. 1977. *Seismic Stratigraphy - Applications to Hydrocarbon Exploration,* Memoir 26, American Association of Petroleum Geologists, Tulsa, OK.

**Pennant-Rea 1994**
Pennant-Rea, R., ed. 1994. "Chainsaw Massacres," *The Economist,* Vol 331, No. 7869, p 39.

**Pethick 1984**
Pethick, J. 1984. *An Introduction to Coastal Geomorphology,* Edward Arnold Publishers, London, UK.

**Pirazzoli 1991**
Pirazzoli, P. A. 1991. *World Atlas of Sea-Level Changes,* Elsevier Scientific Publishers, Amsterdam, The Netherlands.

**Plafker and Kachadoorian 1966**
Plafker, G., and Kachadoorian, R. 1966. "Geologic Effects of the March 1964 Earthquake and Associated Seismic Sea Waves on Kodiak and Nearby Islands, Alaska," Geological Survey Professional Paper 543-D, U.S. Government Printing Office, Washington, DC.

**Pope 1997**
Pope, J. 1997. "Responding to Coastal Erosion and Flooding Damages," *Journal of Coastal Research,* Vol 13, No. 3, pp 704-710.

**Rosen, Brenninkmeyer, and Maybury 1993**
Rosen, P. S., Brenninkmeyer, B. M., and Maybury, L. M. 1993. "Holocene Evolution of Boston Inner Harbor, Massachusetts," *Journal of Coastal Research,* Vol 9, No. 2, pp 363-377.

**Sahagian and Holland 1991**
Sahagian, D. L., and Holland, S. M. 1991. Eustatic Sea-Level Curve Based on a Stable Frame of Reference: Preliminary Results, *Geology,* Vol 19, pp 1208-1212.

**Schwartz 1982**
Schwartz, M. L., ed. 1982. *The Encyclopedia of Beaches and Coastal Environments,* Encyclopedia of Earth Sciences, Volume XV, Hutchinson Ross Publishing Company, Stroudsburg, PA.

**Shalowitz 1962**
Shalowitz, A. L. 1962. *Shore and Sea Boundaries, with Special Reference to the Interpretation and Use of Coast and Geodetic Survey Data.* Vol 1, Pub 10-1, U.S. Department of Commerce, Coast and Geodetic Survey, U.S. Government Printing Office, Washington, DC.

**Shalowitz 1964**
Shalowitz, A. L. 1964. *Shore and Sea Boundaries, with Special Reference to the Interpretation and Use of Coast and Geodetic Survey Data.* Vol 2, Pub 10-1, U.S. Department of Commerce, Coast and Geodetic Survey, U.S. Government Printing Office, Washington, DC.

**Shepard 1973**
Shepard, F. P. 1973. *Submarine Geology,* 3rd ed., Harper & Row, New York, NY.

**Shore Protection Manual 1984**
*Shore Protection Manual.* 1984. 4th ed., 2 Vol, Coastal Engineering Research Center, U.S. Army Engineer Waterways Experiment Station, U.S. Government Printing Office, Washington, DC.

**Simpson and Riehl 1981**
Simpson, R. H., and Riehl, H. 1981. *The Hurricane and Its Impact,* Louisiana State University Press, Baton Rouge, LA.

**Stanley 1986**
Stanley, S. M. 1986. *Earth and Life Through Time,* W. H. Freeman, New York, NY.

**Stauble and Kraus 1993**
Stauble, D. K., and Kraus, N. C. 1993. *Beach Nourishment Engineering and Management Considerations,* Coastlines of the World Series, American Society of Civil Engineers, New York, NY.

**Stauble, Garcia, Kraus, Grosskopf, and Bass 1993**
Stauble, D. K., Garcia, A. W., Kraus, N. C., Grosskopf, W. G., and Bass, G. P. 1993. "Beach Nourishment Project Response and Design Evaluation, Ocean City, Maryland," Technical Report CERC-93-13, U.S. Army Engineer Waterways Experiment Station, Vicksburg, MS.

**Stone and Morgan 1993**
Stone, G. W., and Morgan, J. P. 1993. "Implications for a Constant Rate of Relative Sea-Level Rise During the Last Millennium Along the Northern Gulf of Mexico: Santa Rosa Island, Florida," *Shore and Beach,* Vol 61, No. 4, pp 24-27.

**Strahler 1981**
Strahler, A. N. 1981. *Physical Geology,* Harper & Row, New York, NY.

**Suter and Berryhill 1985**
Suter, J. R., and Berryhill, H. L., Jr. 1985. "Late Quaternary Shelf-Margin Deltas, Northwest Gulf of Mexico," *Bulletin of the American Association of Petroleum Geologists,* Vol 69, No. 1, pp 77-91.

**Tait 1993**
Tait, L. S., compiler. 1993. "The State of the Art of Beach Renourishment," *Proceedings of the 6th Annual National Conference on Beach Preservation Technology,* Florida Shore & Beach Preservation Association, Tallahassee, FL.

**Tannehill 1956**
Tannehill, I. R. 1956. *Hurricanes, Their Nature and History,* 9th Revised ed., Princeton University Press, Princeton, NJ.

**Tanner 1989**
Tanner, W. F. 1989. "New Light on Mean Sea Level Change," *Coastal Research,* Vol 8, No. 4, pp 12-16.

**U.S. Army Corps of Engineers 1995**
U.S. Army Corps of Engineers. 1995. "Coastal Geology," Engineer Manual 1110-2-1810, Washington, DC.

**Whitehill 1968**
Whitehill, W. M. 1968. *Boston, A Topographical History,* 2nd ed. (enlarged), The Belknap Press of Harvard University Press, Cambridge, MA.

**Winkler 1977**
Winkler, C. D. 1977. "Plio-Pleistocene Paleogeography of the Florida Gulf Coast Interpreted from Relic Shorelines," *Transactions Gulf Coast Association of Geological Societies,* Vol 27, pp 409-420.

**Winkler and Howard 1977**
Winkler, C. D., and Howard, J. D. 1977. "Correlation of Tectonically Deformed Shorelines on the Southern Atlantic Coastal Plain," *Geology,* Vol 5, pp 123-127.

**Woodsworth and Wigglesworth 1934**
Woodsworth, J. B., and Wigglesworth, E. 1934. *Geography and Geology of the Region Including Cape Cod, Elizabeth Is., Nantucket, Martha's Vinyard, No Mans Land, and Block Is.,* Memoir 52, Museum of Comparative Zoology, Harvard University, Cambridge, MA.

**Wunsch 1996**
Wunsch, C. 1996. Doherty Lecture: "The Ocean and Climate - Separating Myth from Fact," *Marine Technical Society Journal,* Vol 30, No. 2, pp 65-68.

**Young 1975**
Young, K. 1975. *Geology: The Paradox of Earth and Man,* Houghton Mifflin Co., Boston, MA.

**Young and Hale 1998**
Young, C., and Hale, L. 1998. "Coastal Management: Insurance for the Coastal Zone," *Maritimes,* Vol 40, No. 1, pp 17-19.

## IV-1-9. Acknowledgments

Authors of Chapter IV-1, "Coastal Terminology and Geologic Environments:"

Andrew Morang, Ph.D., Coastal and Hydraulics Laboratory (CHL), Engineer Research and Development Center (ERDC), Vicksburg, Mississippi.
Larry E. Parson, U.S. Army Engineer District, Mobile, Mobile, Alabama.

Reviewers:

Stephan A. Chesser, U.S. Army Engineer District, Portland, Portland, Oregon.
Ronald L. Erickson, U.S. Army Engineer District, Detroit, Detroit, Michigan.
James R. Houston, Ph.D., ERDC.
John H. Lockhart, Jr., Headquarters, U.S. Army Corps of Engineers, Washington, DC, (retired).
Edward P. Meisburger, CHL (retired).
Joan Pope, CHL.
John F. C. Sanda, Headquarters, U.S. Army Corps of Engineers, Washington, DC., (retired).
Orson P. Smith, Ph.D., U.S. Army Engineer District, Alaska, Anchorage, Alaska, (retired).

## Table of Contents

# List of Tables

# List of Figures

# Chapter IV-2
# Coastal Classification and Morphology

## IV-2-1. Introduction

*a.* Since ancient times, men have gone to sea in a variety of vessels to obtain food and to transport cargo and passengers to distant ports. To navigate safely, sailors needed an intimate knowledge of the appearance of the coast from place to place. By the time that systematic study of coastal geology and geomorphology began, there already existed a large body of observational knowledge about seacoasts in many parts of the world and a well-developed nomenclature to portray coastal landforms. Geologists in the 19th and 20th centuries described coastal landforms, examined their origin and development as a function of geologic character, history, and dynamic processes, and devised classification schemes to organize and refine their observations.

*b.* The first part of this chapter discusses the coastal classification of Francis Shepard (1973). The second part describes specific coastal environments found around the United States following Shepard's outline.

## IV-2-2. Coastal Classification

*a. Introduction.* By its very nature, the coast is an incredibly complex and diverse environment, one that may defy organization into neat compartments. Nevertheless, the quest for understanding how shorelines formed and how human activities affect these processes has demanded that classification schemes be devised. Most have grouped coastal areas into classes that have similar features because of having developed in similar geological and environmental settings.

*b. Early classifications.* Many early geologists took a genetic approach to classification and distinguished whether the coast had been primarily affected by rising sea level (submergence), falling sea level (emergence), or both (compound coasts) (Dana 1849; Davis 1896a; Gulliver 1899; Johnson 1919; Suess 1888).

*c. Later classifications.* The best known of the modern classifications are those of Cotton (1952), Inman and Nordstrom (1971), Shepard (1937), with revisions in 1948, 1971 (with Harold Wanless), 1973, and 1976, and Valentin (1952). Except for Inman and Nordstrom (1971), these classifications emphasized onshore and shoreline morphology but did not include conditions of the offshore bottom. This may be a major omission because the submarine shoreface and the shelf are part of the coastal zone. Surprisingly few attempts have been made to classify the continental shelf. Shepard (1948; 1977) and King (1972) discussed continental shelf types, but their classifications are not detailed and contain only a few broadly defined types.

*d. Coastal classification of Francis Shepard.* Possibly the most widely used coastal classification scheme is the one introduced by Shepard in 1937 and modified in later years. It divides the world's coasts into primary coasts - formed mostly by non-marine agents - and secondary coasts - shaped primarily by marine processes. Further subdivisions occur according to which specific agent, terrestrial or marine, had the greatest influence on coastal development. The advantage of Shepard's classification is that it is more detailed than others, allowing most of the world's coasts to be incorporated. Although gradational shore types exist, which are difficult to classify, most coasts show only one dominant influence as the cause of their major characteristics (Shepard 1973). Because of its overall usefulness, Shepard's 1973 classification is reproduced in Table IV-2-1. Specific coasts are discussed in detail in this chapter, approximately following the outline of Shepard's table.

Table IV-2-1
Classification of Coasts (Continued)

| Excerpt from SUBMARINE GEOLOGY, 3rd ed. by Francis P. Shepard. Copyright 1948, 1963, 1973 by Francis P. Shepard. Reprinted by permission of Harper Collins Publishers. | Paragraph No. |
|---|---|
| I. *Primary coasts* Configuration due to nonmarine processes. | |
|    A.  *Land erosion coasts* Shaped by subaerial erosion and partly drowned by postglacial rise of sea level (with or without crustal sinking) or inundated by melting of an ice mass from a coastal valley. | |
|       1.  *Ria coasts (drowned river valleys)* Usually recognized by the relatively shallow water of the estuaries which indent the land. Commonly have V-shaped cross section and a deepening of the axis seaward except where a barrier has built across the estuary mouth. | IV-2-3 |
|       2.  *Dendritic* Pattern resembling an oak leaf due to river erosion in horizontal beds or homogeneous material. | |
|       3.  *Trellis* Due to river erosion in inclined beds of unequal hardness. | |
|         (a) *Drowned glacial erosion coasts* Recognized by being deeply indented with many islands. Charts show deep water (commonly more than 100 m) with a U-shaped cross section of the bays and with much greater depth in the inner bays than near the entrance. Hanging valleys and sides usually parallel and relatively straight, in contrast to the sinuous rias. Almost all glaciated coasts have bays with these characteristics. | IV-2-4 |
|         (b) *Fjord coasts* Comparatively narrow inlets cutting through mountainous coasts. | |
|         (c) *Glacial troughs* Broad indentations, like Cabot Strait and the Gulf of St. Lawrence or the Strait of Juan de Fuca. | |
|       4.  *Drowned karst topography* Embayments with oval-shaped depressions indicative of drowned sinkholes. This uncommon type occurs locally, as along the west side of Florida north of Tarpon Springs, the east side of the Adriatic, and along the Asturias coast of North Spain. | |
|    B.  *Subaerial deposition coasts* | IV-3-3 |
|       1.  *River deposition coasts* Largely due to deposition by rivers extending the shoreline since the slowing of the postglacial sea level rise. | |
|       2.  *Deltaic coasts* | |
|         (a) *Digitate (birdfoot),* the lower Mississippi Delta. | |
|         (b) *Lobate,* western Mississippi Delta, Rhone Delta. | |
|         (c) *Arcuate,* Nile Delta. | |
|         (d) *Cuspate,* Tiber Delta. | |
|         (e) *Partly drowned deltas* with remnant natural levees forming islands. | |
|       3.  *Compound delta coasts* Where a series of deltas have built forward a large segment of the coast, for example, the North Slope of Alaska extending east of Point Barrow to the Mackenzie. | |
|       4.  *Compound alluvial fan coasts straightened by wave erosion.* | |
|       5.  *Glacial deposition coasts* | |
|         (a) *Partially submerged moraines* Usually difficult to recognize without a field study to indicate the glacial origin of the sediments constituting the coastal area. Usually modified by marine erosion and deposition as, for example, Long Island. | |
|         (b) *Partially submerged drumlins* Recognized on topographic maps by the elliptical contours on land and islands with oval shorelines, for example, Boston Harbor and West Ireland (Guilcher 1965). | |
|         (c) *Partially submerged drift features* | IV-2-6 |
|       6.  *Wind deposition coasts* It is usually difficult to ascertain if a coast has actually been built forward by wind deposition, but many coasts consist of dunes with only a narrow bordering sand beach. | |
|         (a) *Dune prograded coasts* Where the steep lee slope of the dune has transgressed over the beach. | |
|         (b) *Dune coasts* Where dunes are bordered by a beach. | |
|         (c) *Fossil dune coasts* Where consolidated dunes (eolianites) form coastal cliffs. | |
|       7.  *Landslide coasts* Recognized by the bulging earth masses at the coast and the landslide topography on land. | |
|    C.  *Volcanic coasts* | IV-2-7 |
|       1.  *Lava-flow coasts* Recognized on charts either by land contours showing cones, by convexities of shoreline, or by conical slopes continuing from land out under the water. Slopes of 10° to 30° common above and below sea level. Found on many oceanic islands. | |
|       2.  *Tephra coasts* Where the volcanic products are fragmental. Roughly convex but much more quickly modified by wave erosion than are lava-flow coasts. | |
|       3.  *Volcanic collapse or explosion coasts* Recognized in aerial photos and on charts by the concavities in the sides of volcanoes. | |
|    D.  *Shaped by diastrophic movements* | |
|       1.  *Fault coasts* Recognized on charts by the continuation of relatively straight steep land slopes beneath the sea. Angular breaks at top and bottom of slope. | |
|         (a) *Fault scarp coasts* For example, northeast side of San Clemente Island, California. | |
|         (b) *Fault trough or rift coasts* For example, Gulf of California and Red Sea, both being interpreted as rifts. | |
|         (c) *Overthrust* No examples recognized but probably exist. | |
|                                      (Continued) | |

| Table IV-2-1 (Concluded) | Paragraph No. |
|---|---|
|     2. *Fold coasts* Difficult to recognize on maps or charts but probably exist.<br>    3. Sedimentary extrusions<br>       (a) *Salt domes* Infrequently emerge as oval-shaped islands. Example: in the Persian Gulf.<br>       (b) *Mud lumps* Small islands due to upthrust of mud in the vicinity of the passes of the Mississippi Delta.<br>  E. *Ice coasts* Various types of glaciers form extensive coasts, especially in Antarctica.<br>II. *Secondary coasts* Shaped primarily by marine agents or by marine organisms. May or may not have been primary coasts before being shaped by the sea.<br>  A. Wave erosion coasts<br>    1. *Wave-straightened cliffs* Bordered by a gently inclined seafloor, in contrast to the steep inclines off fault coasts.<br>       (a) *Cut in homogeneous materials.*<br>       (b) *Hogback strike coasts* Where hard layers of folded rocks have a strike roughly parallel to the coast so that erosion forms a straight shoreline.<br>       (c) *Fault-line coasts* Where an old eroded fault brings a hard layer to the surface, allowing wave erosion to remove the soft material from one side, leaving a straight coast.<br>       (d) *Elevated wave-cut bench coasts* Where the cliff and wave-cut bench have been somewhat elevated by recent diastrophism above the level of present-day wave erosion.<br>       (e) *Depressed wave-cut bench coasts* Where the wave-cut bench has been somewhat depressed by recent diastrophism so that it is largely below wave action and the wave-cut cliff plunges below sea level.<br>    2. *Made irregular by wave erosion* Unlike ria coasts in that the embayments do not extend deeply into the land. *Dip coasts* Where alternating hard and soft layers intersect the coast at an angle; cannot always be distinguished from trellis coasts.<br>       (a) *Heterogeneous formation coasts* Where wave erosion has cut back the weaker zones, leaving great irregularities.<br>  B. *Marine deposition coasts* Coasts prograded by waves and currents.<br>    1. Barrier coasts.<br>       (a) *Barrier beaches* Single ridges.<br>       (b) *Barrier islands* Multiple ridges, dunes, and overwash flats.<br>       (c) *Barrier spits* Connected to mainland.<br>       (d) *Bay barriers* Sand spits that have completely blocked bays.<br>       (e) *Overwash fans* Lagoonward extension of barriers due to storm surges.<br>    2. *Cuspate forelands* Large projecting points with cusp shape. Examples include Cape Hatteras and Cape Canaveral.<br>    3. *Beach plains* Sand plains differing from barriers by having no lagoon inside.<br>    4. *Mud flats or salt marshes* Formed along deltaic or other low coasts where gradient offshore is too small to allow breaking waves.<br>  C. *Coasts* built by organisms<br>    1. *Coral reef coasts* Include reefs built by coral or algae. Common in tropics. Ordinarily, reefs fringing the shore and rampart beaches are found inside piled up by the waves.<br>       (a) *Fringing reef coasts* Reefs that have built out the coast.<br>       (b) *Barrier reef coasts* Reefs separated from the coast by a lagoon.<br>       (c) *Atolls* Coral islands surrounding a lagoon.<br>       (d) *Elevated reef coasts* Where the reefs form steps or plateaus directly above the coast.<br>    2. *Serpulid reef coasts* Small stretches of coast may be built out by the cementing of serpulid worm tubes onto the rocks or beaches along the shore. Also found mostly in tropics.<br>    3. *Oyster reef coasts* Where oyster reefs have built along the shore and the shells have been thrown up by the waves as a rampart.<br>    4. *Mangrove coasts* Where mangrove plants have rooted in the shallow water of bays, and sediments around their roots have been built up to sea level, thus extending the coast. Also a tropical and subtropical development.<br>    5. *Marsh grass coasts* In protected areas where salt marsh grass can grow out into the shallow sea and, like the mangroves, collect sediment that extends the land. Most of these coasts could also be classified as mud flats or salt marshes. | IV-2-8<br><br><br><br><br><br><br><br><br><br><br><br><br><br><br><br>IV-2-9<br><br><br><br><br><br><br><br><br>IV-2-11<br><br>IV-2-12 |

*e. Classification schemes for specific environments.*

(1) River systems. Coleman and Wright (1971) developed a detailed classification for rivers and deltas.

(2) Great Lakes of North America. The Great Lakes have unique characteristics that set them apart from oceanic coastlines. One of the most comprehensive attempts to include these factors in a classification scheme was developed by Herdendorf (1988). It was applied to the Canadian lakes by Bowes (1989). A

simpler scheme has been used by the International Joint Commission as a basis for studies of shoreline erosion (Stewart and Pope 1992).

## IV-2-3. Drowned River Coasts - Estuaries[1]

*a. Introduction.* An enormous amount of technical literature is devoted to the chemistry and biology of estuaries. In recent years, much research has been devoted to estuarine pollution and the resulting damage to fish and animal habitat. For example, the famous oyster harvesting in Chesapeake Bay has been almost ruined in the last 30 years because of overfishing, urban runoff, and industrial pollution. As a result, the unique way of life of the Chesapeake oystermen, who still use sailing vessels, may be at an end. Possibly because most attention has centered on the biological and commercial aspects of estuaries, our geological understanding of them is still rudimentary (Nichols and Biggs 1985). The estuarine environment can be defined as the complex of lagoon-bay-inlet-tidal flat and marsh that make up 80 to 90 percent of the U.S. Atlantic and Gulf coasts (Emery 1967). Clearly it is vital that we gain a better understanding of their sedimentary characteristics and dynamics.

*b. Literature.* Only the briefest introduction to estuarine processes and sediments can be presented in this chapter. The purpose of this section is to introduce estuarine classification, regional setting, and geology. The reader is referred to Nichols and Biggs (1985) for an overview of the geology and chemistry of estuaries and for an extensive bibliography. Other general works include Dyer (1979), Nelson (1972), and Russell (1967). Cohesive sediment dynamics are covered in Metha (1986), and the physics of estuaries is covered in van de Kreeke (1986). Research from the 1950's and 1960's is reviewed in Lauff (1967).

*c. Classification.* Many attempts have been made to define and classify estuaries using geomorphology, hydrography, salinity, sedimentation, and ecosystem parameters (reviewed in Hume and Herdendorf (1988)). A geologically based definition, which accounts for sediment supply pathways, is used in this text.

*d. Definitions.* Estuaries are confined bodies of water that occupy the drowned valleys of rivers that are not currently building open-coast deltas. The most common definition of an estuary describes it as a body of water where "...seawater is measurably diluted with fresh water derived from land drainage" (Pritchard 1967). Therefore, estuaries would include bodies of water where salinity ranges from 0.1 ‰ (parts per thousand) to about 35 ‰ (Figure IV-2-1). However, this chemical-based definition does not adequately restrict estuaries to the setting of river mouths, and allows, for example, lagoons behind barrier islands to be included. Dalrymple, Zaitlin, and Boyd (1992) felt that the interaction between river and marine processes was an attribute essential to all true estuaries. Therefore, they proposed a new geologically based definition of estuary as:

> ...the seaward portion of a drowned valley system which receives sediment from both fluvial and marine sources and which contains facies influenced by tide, wave, and fluvial processes. The estuary is considered to extend from the landward limit of tidal facies at its head to the seaward limit of coastal facies at its mouth.

These limits are schematically shown in Figure IV-2-1.

*e. Time relationships and evolution.*

(1) Estuaries, like other coastal systems, are ephemeral. River mouths undergo continuous geological evolution, of which estuaries represent one phase of a continuum (Figure IV-2-2). During a period of high

---

[1] Material in this section has been condensed from Dalrymple, Zaitlin, and Boyd (1992).

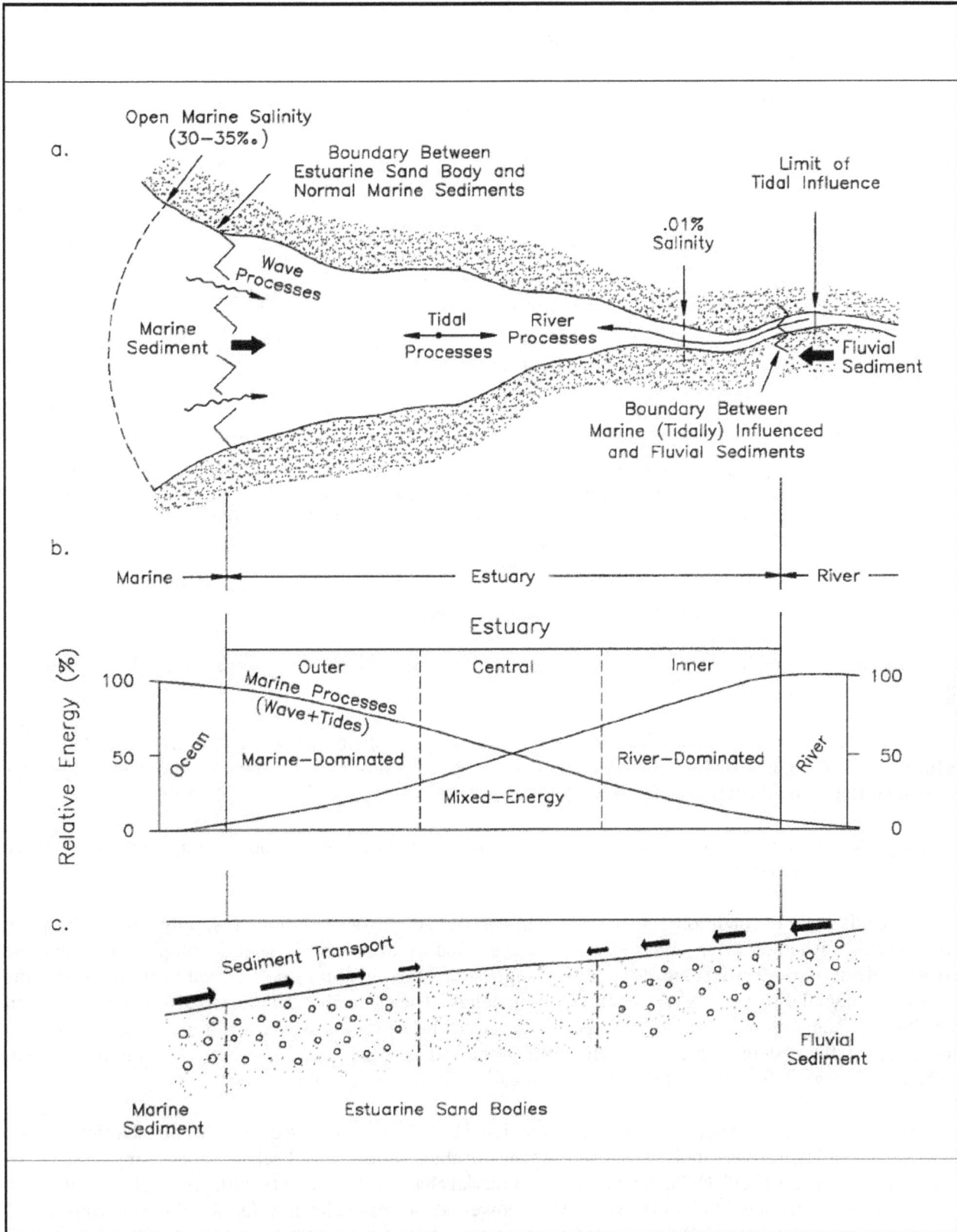

Figure IV-2-1. (a) Plan view of distribution of energy and physical processes in estuaries; (b) Schematic definition of estuary according to Dalyrmple, Zaitlin, and Boyd (1992); (c) Time-averaged sediment transport paths

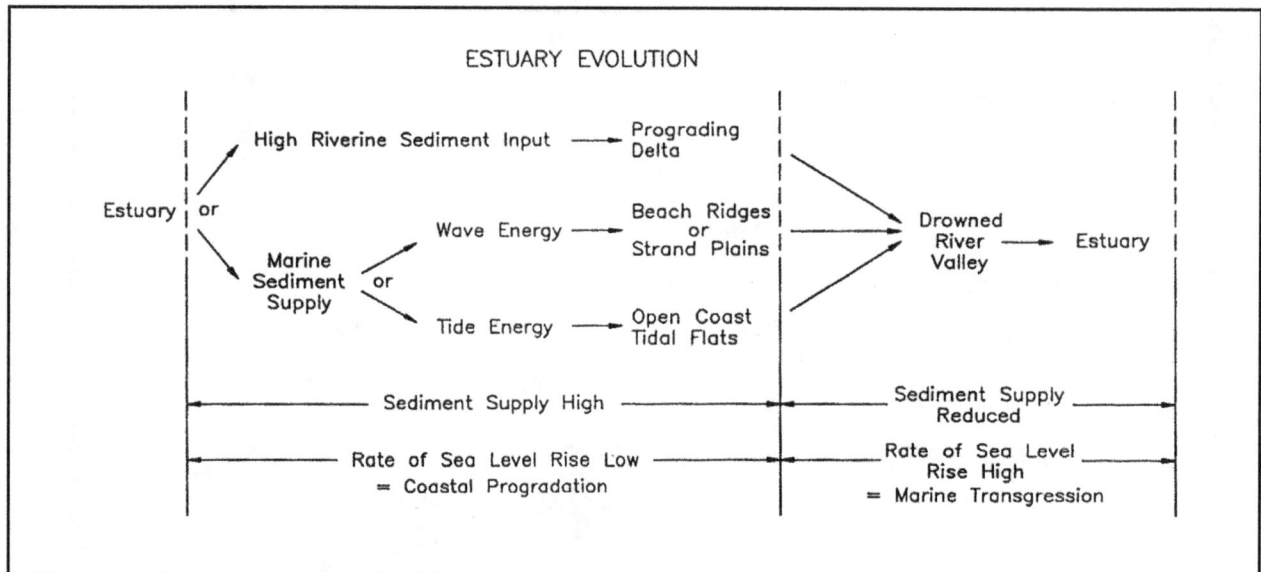

Figure IV-2-2.  Estuary/prograding coast evolution.  Estuaries are part of a continuum of coastal landforms. With high riverine or marine sediment supply, the shore prograde (left half of figure).  Later, if sediment supply is reduced, the river valley is drowned, resulting in an estuary (right half of figure)

sediment supply and low rate of sea level rise, an estuary is gradually filled.  Three coastal forms may result, depending on the balance between riverine input and marine sediment supply:

(a)  If the sediment is supplied by a river, a delta is formed, which, as it grows, prograde out into the open sea (left side of Figure IV-2-3).

(b)  If, instead, most sediment is delivered to the area by marine processes, a straight, prograding coast is formed.  This might be in the form of beach ridges or strand plains if wave energy is dominant, or as open-coast tidal flats if tidal energy is dominant.

(c)  Later, if sea level rises at a higher rate, then the river valley may be flooded, forming a new estuary (right side of Figure IV-2-3).

(2)  Under some conditions, such as when sea level rise and sediment supply are in balance, distinguishing whether a river mouth should be classified as an estuary or as a developing delta may be difficult.  Dalrymple, Zaitlin, and Boyd (1992) suggest that the direct transport of bed material may be the most fundamental difference between estuaries and deltas.  They state that the presence of tight meanders in the channels suggests that bed-load transport is landward in the region seaward of the meanders and, as a consequence, the system is an estuary.  However, if the channels are essentially straight as far as the coast, bed load is seaward throughout the system and it can be defined as a delta.

(3)  Fluvial systems are controlled by their erosional base level and the sediment supply.  During periods of lowered sea level, rivers incise the lower reaches of their valleys and discharge increasing amounts of sediment out onto the shelf.  Deltas accumulate and fluvial channels are cut, dissecting parts of the delta plain (described in greater detail in Part IV-3-3).  At the lowest stands of sea level, estuaries almost disappear and are confined to river valleys (Baeteman 1994).  When sea level rises again, the valleys are flooded and the estuaries reappear.

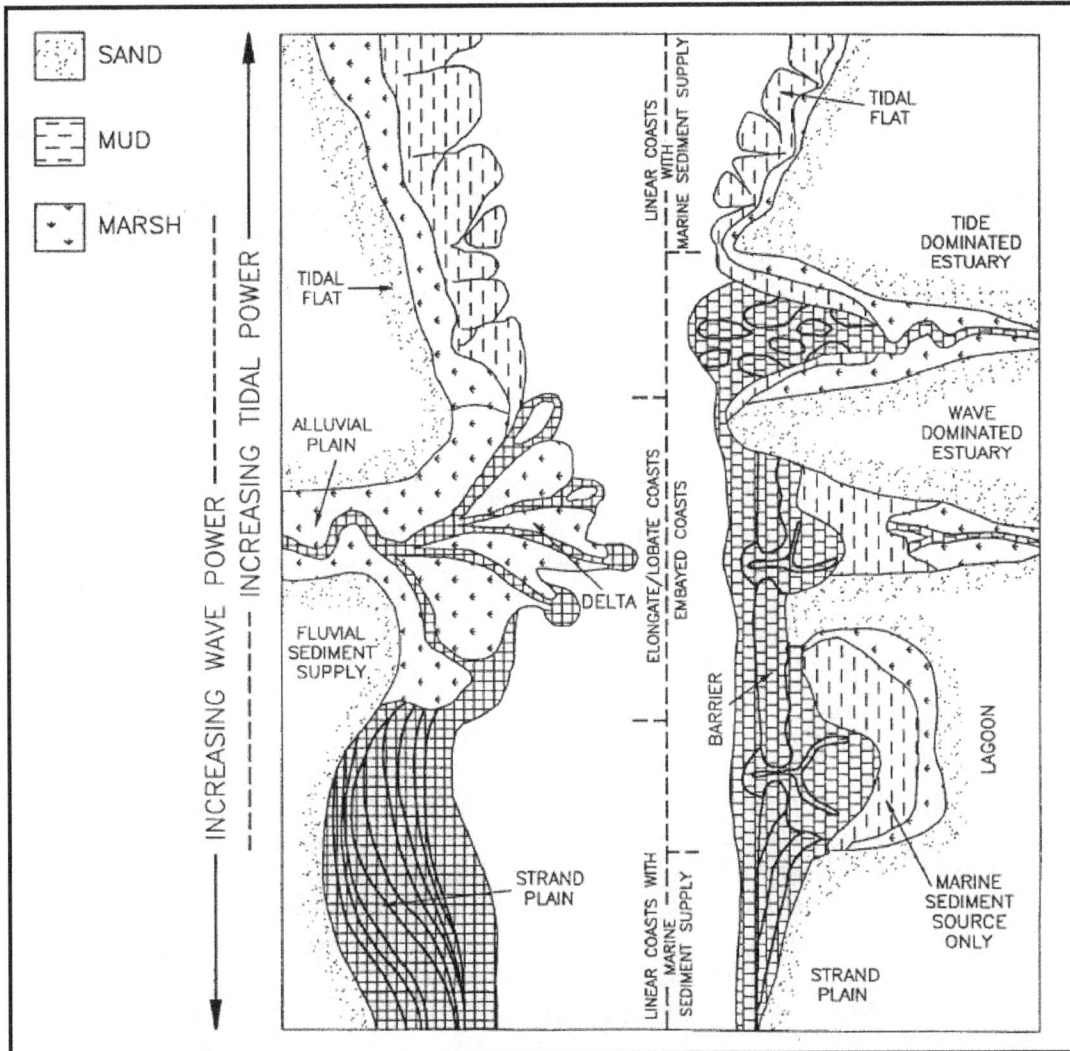

Figure IV-2-3. Estuary evolution, based on changes in wave or tidal power. The left half of the figure shows a prograding coast that results during times of high sediment supply. The right side shows how estuaries develop during reduced sediment supply (also refer to Figure IV-2-2). Adapted from Boyd, Dalrymple, and Zaitlin (1992)

*f. Overall geomorphic characteristics.* The geologic definition of estuary implies that sediment supply does not keep pace with the local sea level rise; as a result, estuaries become sinks for terrestrial and marine sediment. Sedimentation is the result of the interaction of wave, tide, and riverine forces. All estuaries, regardless of whether they are wave- or tide-dominated, can be divided into three zones (Figure IV-2-1):

(1) The *outer zone* is dominated by marine processes (wave and tidal currents). Because of currents, coarse sediment tends to move up into the mouth of the estuary.

(2) The *central zone* is characterized by relatively low energy, where wave and tidal currents are balanced over the long term by river currents. The central zone is an area of net convergence of sediment and usually contains the finest-grained bed load present in the estuary.

(3) The *inner zone* is river-dominated and extends upriver to the limit of tidal influence. The long-term (averaged over years) bed load transport in this region is seaward.

*g.   Energy factors and sedimentary structures.*

(1)  Wave-dominated estuaries.

(a)  This type of estuary is characterized by high wave energy compared to tidal influence.  Waves cause sediment to move alongshore and onshore into the mouth of the estuary, forming sandbars or subaerial barriers and spits (Figure IV-2-4a).  The barrier prevents most of the wave energy from entering the central basin.  In areas of low tide range and small tidal prism, tidal currents may not be able to maintain the inlet, and storm breaches tend to close during fair weather, forming enclosed coastal ponds.  Sediment type is well-distributed into three zones, based on the variation of total energy: coarse sediment near the mouth, fine in the central basin, and coarse at the estuary head.  A marine sand body forms in the high wave energy zone at the mouth.  This unit is composed of barrier and inlet facies, and, if there is moderate tide energy, sand deposited in flood-tide deltas (Hayes 1980).

(b)  At the head of the estuary, the river deposits sand and gravel, forming a bay-head delta.  If there is an open-water lagoon in the central basin, silts and fine-grained organic muds accumulate at the toe of the bay-head delta.  This results in the formation of a prodelta similar to the ones found at the base of open-coast deltas (deltaic terms and structures are discussed in Part IV-3).  Estuaries that are shallow or have nearly filled may not have an open lagoon.  Instead, they may be covered by extensive salt marshes crossed by tidal channels.

(2)  Tide-dominated estuaries.

(a)  Tide current energy is greater than wave energy at the mouth of tide-dominated estuaries, resulting in the development of elongate sandbars (Figure IV-2-4b).  The bars dissipate wave energy, helping protect the inner portions of the estuary.  However, in funnel-shaped estuaries, the incoming flood tide is progressively compressed into a decreasing cross-sectional area as it moves up the bay.  As a result, the velocity of the tide increases until the effects of the amplification caused by convergence are balanced by frictional dissipation.  The velocity-amplification behavior is known as *hypersynchronos* (Nichols and Biggs 1985).  Because of friction, the tidal energy decreases beyond a certain distance in the estuary, eventually becoming zero.

(b)  As in wave-dominated estuaries, riverine energy also decreases downriver from the river mouth.  The zone where tide and river energy are equal is sometimes called a balance point and is the location of minimum total energy.  Because the total-energy minimum is typically not as low as the minimum found in wave-dominated estuaries, tide-dominated estuaries do not display as clear a zonation of sediment facies.  Sands are found along the tidal channels, while muddy sediments accumulate in the tidal flats and marshes along the sides of the estuary (Figure IV-2-4b).  In the central, low-energy zone, the main tidal-fluvial channel consistently displays a sinuous, meandering shape.  Here, the channel develops alternate bars at the banks and, sometimes, in mid-channel.

(c)  A bay-head delta is usually not present in the river-dominated portion of tidally dominated estuaries.  Instead, the river channel merges directly into a single or a series of tidal channels that eventually reach the sea.

(3)  Estuarine variability.

(a)  Wave-to-tide transition.  As tide energy increases relative to wave energy, the barrier system at the mouth of the estuary becomes progressively more dissected by tidal inlets, and elongate sandbars form along

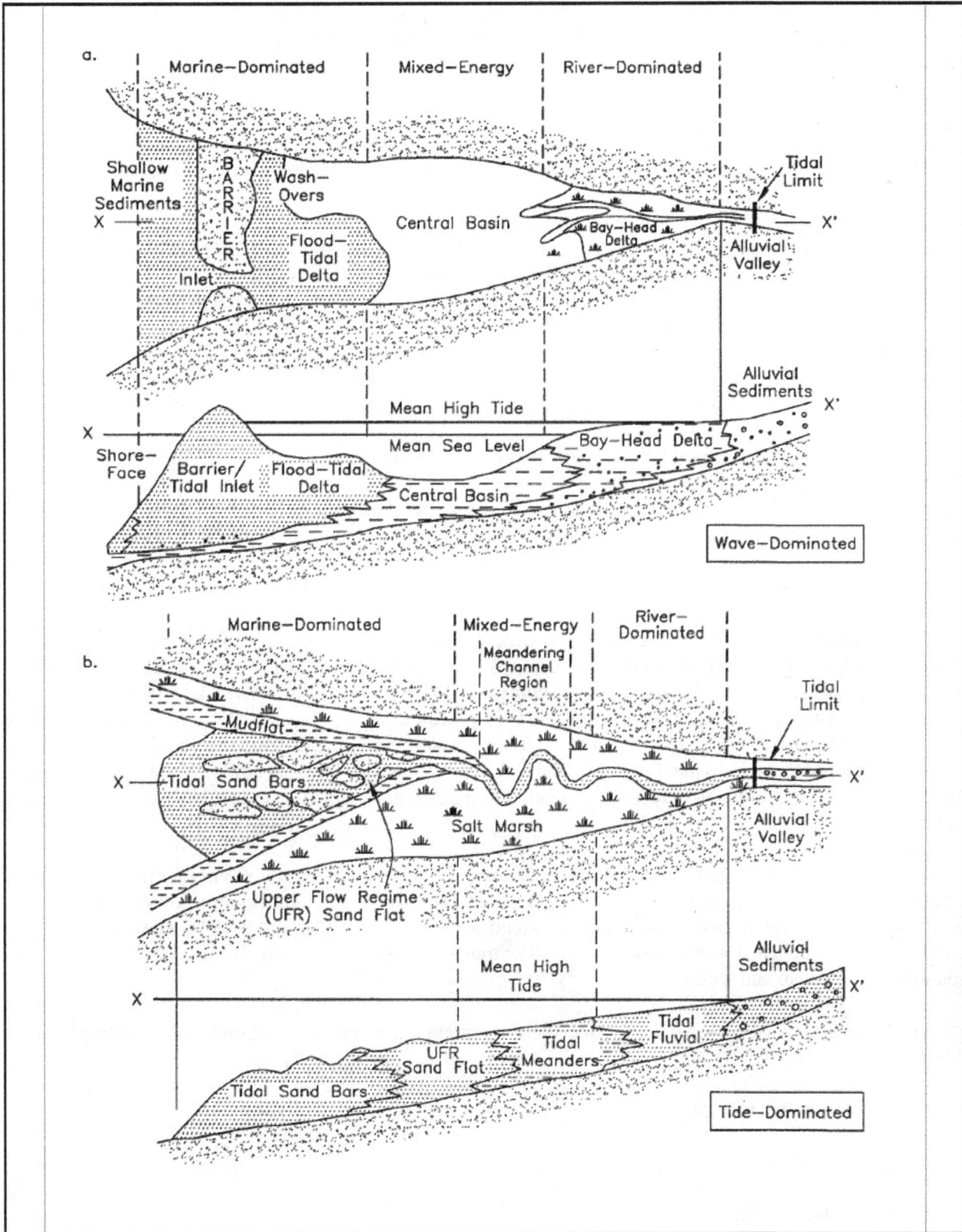

Figure IV-2-4. Morphologic models of (a) wave-dominated and (b) tide-dominated estuaries (adapted from Dalyrmple, Zaitlin, and Boyd (1992)). Wave-dominated estuaries are common along the mid-Atlantic coast of the United States. Tide-dominated estuaries are found in Maine, Massachusetts, and the mid-Atlantic Bight

the margins of the tidal channels. As energy levels increase in the central, mixed-energy part of the estuary, marine sand is transported further up into the estuary, and the muddy central basin is replaced by sandy tidal channels flanked by marshes.

(b) Effects of tide range. The inner end of an estuary has been defined as the limit of detectable tidal influence. Therefore, the gradient of the coastal zone and the tide range have a great influence on the length of estuaries (Dalrymple, Zaitlin, and Boyd 1992). Estuaries become longer as gradient decreases and tide range increases.

(c) Influence of valley shape. The shape of the flooded valley and the pre-existing geology also control the size of the estuary and the nature of sediment deposition. This is particularly evident during the early phases of estuary infilling, before erosion and deposition have modified the inherited geology. For example, tidal-wave amplification is less likely to occur in irregular valleys (Nichols and Biggs 1985). The resulting estuaries are more likely to become wave-dominated. Chesapeake Bay, with its extensive system of tributary valleys, is an example of this type. In contrast, estuaries that initially or later have developed a funnel shape are more likely to be tide-dominated and hypersynchronous (for example, the Gironde Estuary of France.)

(d) Geologic setting. Coastal plain gradient, part of the overall plate tectonic setting, is one factor that determines estuary volume. Sea level rise over a flat coastal plain on a passive margin like the Gulf of Mexico creates long estuaries with large volume. An equivalent rise on a steep, active-margin coast like the U.S. Pacific coast will result in a much smaller estuary volume (Boyd, Dalrymple, and Zaitlin 1992).

*h. Estuarine sediments.*

(1) Because estuaries occupy drowned river valleys, they function as sinks for enormous volumes of sediment. Estuarine sediments are derived from various sources including rivers, the continental shelf, local erosion, and biological activity. Sedimentation is controlled by tides, river flow, waves, and meteorology. The lower-energy conditions of estuaries, as opposed to those found on open coasts, allow for the deposition of fine-grained silts, muds, clays, and biogenic materials. Estuarine sediments are typically soft and tend to be deposited on smooth surfaces that limit turbulence of the moving water. When allowed to accumulate, these materials consolidate and undergo various chemical and organic changes, eventually forming cohesive sediments.

(2) The shores of estuaries and certain open-water coasts in low-energy environments (e.g., coastal Louisiana, Surinam, Bangladesh, and Indonesia) are characterized as having smooth, low-sloping profiles with turbid water occurring along the shore and extending well offshore (Suhayda 1984). These areas usually exhibit low and vegetated backshores and mud flats that are exposed at low tide. These conditions are also found in Chesapeake and Delaware Bays.

(3) Nichols and Biggs (1985) describe the movement of estuarine sediments as consisting of four processes:

(a) Erosion of bed material.

(b) Transportation.

(c) Deposition on the bed.

(d) Consolidation of deposited sediment.

These processes strongly depend on estuarine flow dynamics and sediment particle properties. The properties most important for cohesive sediments are interparticle bonding and chemical behavior because these parameters make cohesive sediment respond quite differently to hydrodynamic forces than do noncohesive sediments. Due to the cohesive bonding, consolidated materials (clays and silts) require higher forces to mobilize, making them more resistant to erosion. However, once cohesive sediment is eroded, fine-grained clays and silts can be transported at much lower velocity than is required for the initiation of erosion.

## IV-2-4. Drowned Glacial Erosion Coasts

*a. Introduction.* During the Pleistocene epoch, massive continental glaciers, similar to the present Antarctic and Greenland ice caps, covered broad parts of the continents. The glaciers waxed and waned in cycles, probably because of climatic variations, causing vast changes to the morphology of coastal regions in the northern latitudes. As a result, glacially modified features dominate the northern coasts and continental shelves, although in many areas marine processes have reworked the shore and substantially modified the glacial imprint.

*b. Erosion and sediment production.* Because glacial ice is studded with rock fragments plucked from the underlying rock, a moving glacier performs like a giant rasp that scours the underlying land surfaces. This process, driven by the great size and weight of the ice sheets, caused enormous erosion and modification to thousands of square kilometers during the Pleistocene epoch.

(1) Fjords. The most spectacular erosion forms are drowned glacial valleys known, as *fjords*, that indent the coasts of Alaska, Norway, Chile, Siberia, Greenland, and Canada (Figure IV-2-5). The over-deepened valleys were invaded by the sea as sea level rose during the Holocene epoch. Today, fjords retain the typical U-shaped profile that is also seen in formerly glaciated mountain valleys. Fjords and other drowned glacial erosion features give Maine a spectacular, rugged coastline (Figure IV-2-6).

(2) Depositional features. As a glacier moves, huge amounts of sediment are incorporated into the moving mass. When the ice melts at the glacial front's furthest advance, the sediment load is dropped. Although the major part of the transported material is dumped in the form of a terminal moraine, some sediments are carried further downstream by meltwater streams (Reineck and Singh 1980). The result is a number of distinctive geomorphic features such as drumlins, fjords, moraines, and outwash plains that may appear along the coast or on the submerged continental shelf (Figure IV-2-7). During submergence by the transgressing sea, the features may be modified to such a degree that their glacial origin is lost. This is especially true of outwash, which is easily reworked by marine processes. The original town of Boston was settled on drumlins in the 1600's (Figure IV-1-18), and the islands in Boston Harbor are reworked drumlins (Figure IV-2-8). Nantucket Island, Block Island, and Long Island are partially submerged moraines that have been extensively reworked (Figure IV-2-9).

*c. Variability.* Glaciated coasts typically display a greater variety of geomorphic forms than are seen in warmer latitudes. The forms include purely glacial, glacio-fluvial, and marine types (Fitzgerald and Rosen 1987). Complexity is added by marine reworking, which can produce barriers, shoals, gravel shores, and steep-cliffed shores. Because of the steep slopes of many glacial coasts, slumping and turbidity flow are major erosive agents. In northern latitudes, the shallow seafloor is gouged by icebergs. In summary, classification of shores in drowned glacial environments can be a major challenge because of the complicated geological history and the large diversity of structures.

*d. Atlantic coast.* A fundamental division of coastal characteristics occurs along the Atlantic coast of North America due to the presence of glacial moraines. The Wisconsin terminal moraine formed a prominent series of islands (i.e., Long Island, Block Island, Nantucket, and Martha's Vineyard) and offshore banks

Figure IV-2-5. Glacial fjord coast: Alaska (Lake George, with Surprise glacier in the background)

(Georges and Nova Scotian Banks). South of the moraine, the topography is flatter and more regular, except for piedmont streams, which intersect the coastal plain.

*e. Offshore geology.* Coasts altered by glaciers usually have offshore regions that are highly dissected by relict drainage systems. These sinuous stream channels display highly irregular and varied topography and are composed of sediment types ranging from outwash sand and gravels to till. Note that relict stream channels are also found on continental shelves in temperate climates, for example, off the coast of Texas (Suter and Berryhill 1985). Channels from both temperate and colder environments, and the associated shelf-margin deltas, were formed during late Quaternary lowstands of sea level and are indicators of the position of ancient coastlines.

## IV-2-5. River Deposition Coasts - Deltas

Deltas are discussed in Part IV-3. Because energy factors and deltaic structures are intimately linked, morphology and river mouth hydrodynamics are discussed together.

## IV-2-6. Wind Deposition Coasts - Dunes

*a. Introduction.* Sand dunes are common features along sandy coastlines around the world. The only climatic zone lacking extensive coastal dunes is the frozen Arctic and Antarctic (although thin dune sheets on the coast of McMurdo Sound, Antarctica, have been described by Nichols (1968)). Sediment supply is probably the most crucial factor controlling growth of dunes; while there is rarely a lack of wind in most coastal areas, some lack sufficient loose sediment (Carter 1988). Dunes serve multiple valuable purposes:

**Figure IV-2-6. Drowned glacial erosion coast: Maine (Potts Point, South Harpswell, near Brunswick, July 1994). Rock headlands and ridges run southwest into the Gulf of Maine**

as recreational areas, as habitat for various species of birds, as shore protection, and as temporary sources and sinks of sand in the coastal environment. Although dunes are found along many sandy coasts, they are finite resources and need to be protected and preserved. The seminal work on dunes is Brigadier R.A. Bagnold's *The Physics of Blown Sand and Desert Dunes* (Bagnold 1941). More than 50 years after its publication, this book continues to be cited because of its sound basis on the laws of physics and its readability. Part III-4 of the *Coastal Engineering Manual* reviews the physics of wind-blown sediment transport and presents methods that can be used to estimate transport volumes.

*b. Origin of dunes.* Many large dune fields are believed to have originated when sea level was lower and sediment supply was greater (Carter 1988). Many are on prograding shorelines, although shoreline advance does not seem to be a requirement for dune formation. In northwest Europe, most of the dunes formed from shelf debris that moved onshore during the late Pleistocene and early Holocene by rising sea level. Dune-building phases have been interrupted by periods of relative stability, marked by the formation of soils. The dunes at Plum Island, Massachusetts, may have formed after 1600 (Goldsmith 1985).

*c. Sediment sources.* The normally dry backshore of sandy beaches may be the most common source of dune sands. A flat or low-relief area inland of the coastline is needed to accommodate the dunes, and there must be predominant onshore or alongshore winds for at least part of the year. To move sand from the beach to the dunes, wind speed must exceed a threshold velocity for the particular size of sand available. If the sand is damp or if the grains must move up a slope, the velocities required for sediment transport are greatly increased. The foreshore of the beach can also be a source of sand if it dries between tidal cycles. This is especially true in areas where there is only one high tide per day (diurnal), allowing a greater amount of time

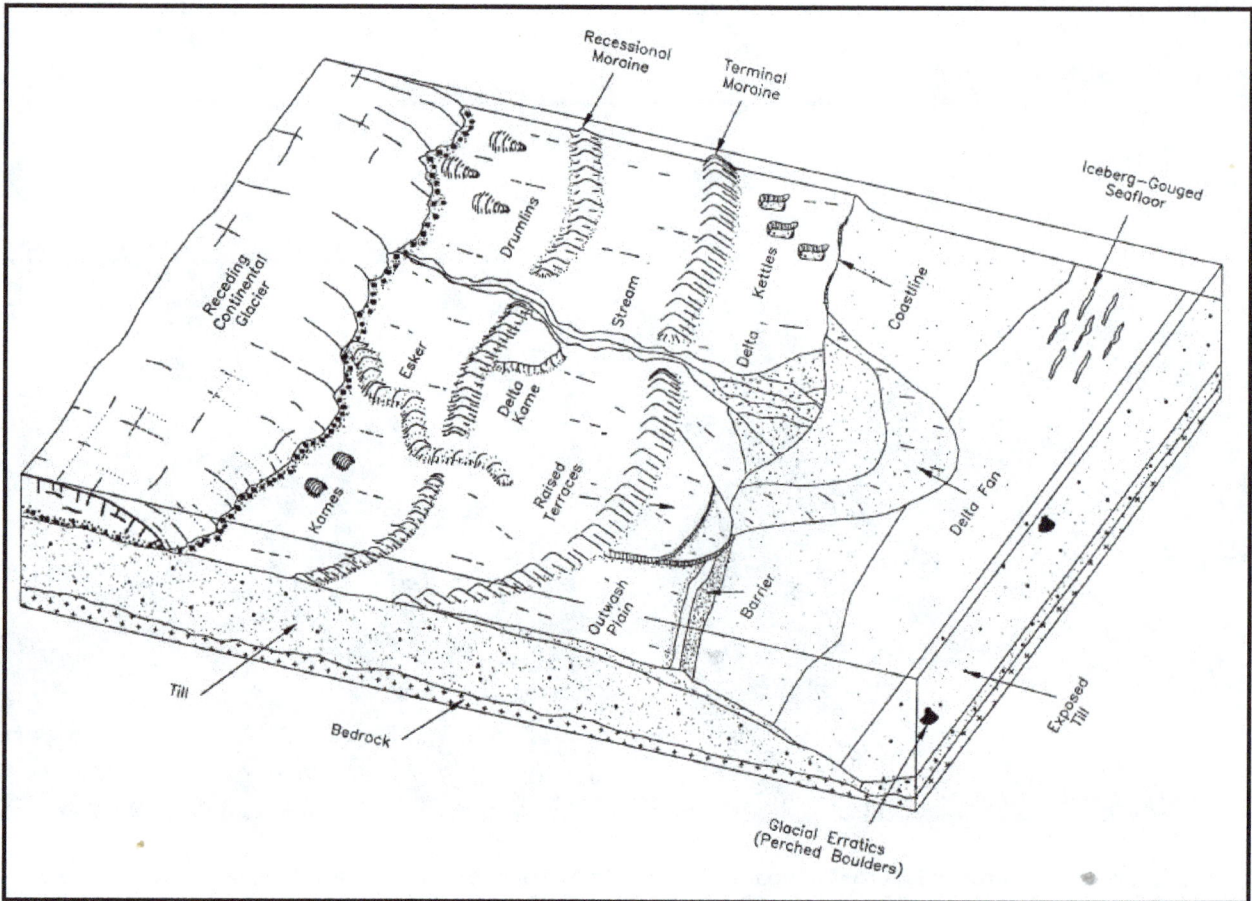

**Figure IV-2-7. Typical glacial depositional structures**

for the foreshore to dry between inundations. Sand storage in dunes must be estimated as one component of sediment budget calculations (EM 1110-2-1502).

*d. Modification and stability.* Most dunes show evidence of post-depositional modifications. These include:

(1) Physical changes - slumping, compaction. Sand grains become rounded, frosted, and better sorted.

(2) Chemical alterations - oxidation, leaching, calcification. (The latter can solidify a dune, making it much more resistant to erosion.)

(3) Biological effects - reactivation, humification, soil formation.

The stability of dunes varies greatly, usually depending on vegetation cover. Dunes in arid climates are often not vegetated and are mobile. However, coastal dunes are normally vegetated by plant species adapted to the harsh coastal environment (Figure IV-2-10). Many dune grasses have long roots, rhizomes, and runners that help hold sand in place. In addition, dense vegetation displaces the aerodynamic boundary of the wind velocity profile upwards. This process produces a net downward momentum flux, promoting sediment trapping (Carter 1988).

Figure IV-2-8. Islands in Boston Harbor, Massachusetts (August 1988 - view looking south).   These are glacial drumlins that have been extensively reworked by contemporary marine processes.  The town of Winthrop (with a tall water tower) is on the drumlin in the center.  Deer Island, in the harbor, is attached to Winthrop with a causeway

Figure IV-2-9.  Glacial till bluffs just west of Montauk Point, Long Island, New York, facing the Atlantic Ocean (March 1998).  As the bluffs erode, the fine material is carried away by waves, leaving a lag of boulders and cobble on the shoreface.  In this area, the seafloor offshore is also covered with gravel, cobble, and coarse sand.  Sand from the Montauk bluffs is carried by littoral currents to the west, where it nourishes the barrier beaches of Long Island's south shore

**Figure IV-2-10. Partly vegetated coastal sand dunes. Rhizomes help hold sand in place and colonize the dune grasses. Eastern Alabama on the Morgan peninsula east of the mouth of Mobile Bay (March 1991). This area was devastated by Hurricane Frederic in 1979 and is slowly recovering. Commercial construction now threatens these dunes**

*e. Dune vegetation.* American beach grass (*Ammophilia breviligulata*) is the most common dune plant in the United States northeast and on the west coast. Along the Gulf coast and the southeast, sea oats (*Uniola paniculata*) is the most abundant species on the dunes. Both plants are remarkably adapted to this environment and are essential to dune stability. They are tolerant of salt spray and occasional inundation by salt water. Growth is stimulated by sand burial, which occurs frequently on the dune. The plant leaves help trap sand on the dune by raising the laminar boundary layer of the wind velocity profile and causing eolian deposition. Regrowth occurs even after rapid deposition of sand up to 1 m thick. Plant growth is by seed and by rhizome extension. Rhizome extension allows rapid plant distribution to help stabilize the surface of the dune, and allows upward growth of the plant to keep pace with sand deposition. Vehicle and foot traffic that damage the vegetation can greatly diminish dune stability.

*f. Dune fauna.* Dunes may appear to be harsh and inhospitable environments, but they are hosts to many species of animals. A wide variety of invertebrates are present, including species as large as crabs. Numerous shorebirds and upland birds use the dune zone for foraging and nesting. Other vertebrates, such as rabbit, fox, deer, etc., frequent the dune in search of food.

*g. Classification.* Dunes can be described or classified on the basis of physical description (external form and internal bedding) or genetic origin (mode of formation). Smith (1954) devised a descriptive classification system that has been widely used. It established the following types (Figure IV-2-11):

(1) Foredunes. Mounds or ridges directly by the beach. Serve as storm buffer.

(2) Parabolic dunes. Arcuate sand ridges with the concave portion facing the beach. Rare; often form downwind of pools or damp areas.

(3) Barchan dunes. Crescent-shaped dunes with the extremities (horns) extending downwind (caused by the horns migrating more rapidly than the central portions). Sometimes show incomplete sand cover moving over a nonerodible pavement.

(4) Transverse dune ridges. Ridges oriented perpendicular or oblique to the dominant winds. Their form is asymmetrical with steep lee and gentle upwind slopes.

(5) Longitudinal (seif) dunes. Dune ridges elongated parallel to the wind direction and symmetrical in profile. Occur in groups over wide areas; feature sinuous crestlines.

(6) Blowouts. Hollows or troughs cut into dunes; may be caused when vehicles or pedestrians damage vegetation (Figure IV-2-12).

(7) Attached dunes. Formations of sand that have accumulated around obstacles such as rocks (Figure IV-2-13).

*h. Shoreline protection.* In many areas, dunes serve a vital role in protecting inland areas from storm surges and wave attack. As a result, many communities require that buildings be erected behind the dunes or beyond a certain distance (a *setback*) from an established coastline. Unfortunately, the protection is ephemeral because severe storms can overtop and erode the dunes, and changes in sediment supply or local wind patterns (sometimes caused by structures and urban development) can leave them sand-starved. If dunes are cut for roads or for walkways, they become particularly vulnerable to erosion. However, compared with hard structures like seawalls, many communities prefer the protection provided by dunes because of aesthetic considerations.

*i. Dune restoration.* Historically, sand dunes have suffered from human pressure, and many dune systems have been irreversibly altered by man, both by accident or design. Many coastal areas in Europe, North America, Australia, and South Africa, which had once-stable forested dunes, have been deforested. The early settlers to New England in the 1600's severely damaged the dune vegetation almost immediately upon their arrival by overgrazing and farming. Dune rebuilding and revegetation have had a long history, mostly unsuccessful (Goldsmith 1985). Recent restoration practices have been more effective (Knutson 1976, 1978; Woodhouse 1978). The two main methods for rebuilding or creating coastal dunes are artificial planting and erecting sand fences. Hotta, Kraus, and Horikawa (1991) review sand fence performance.

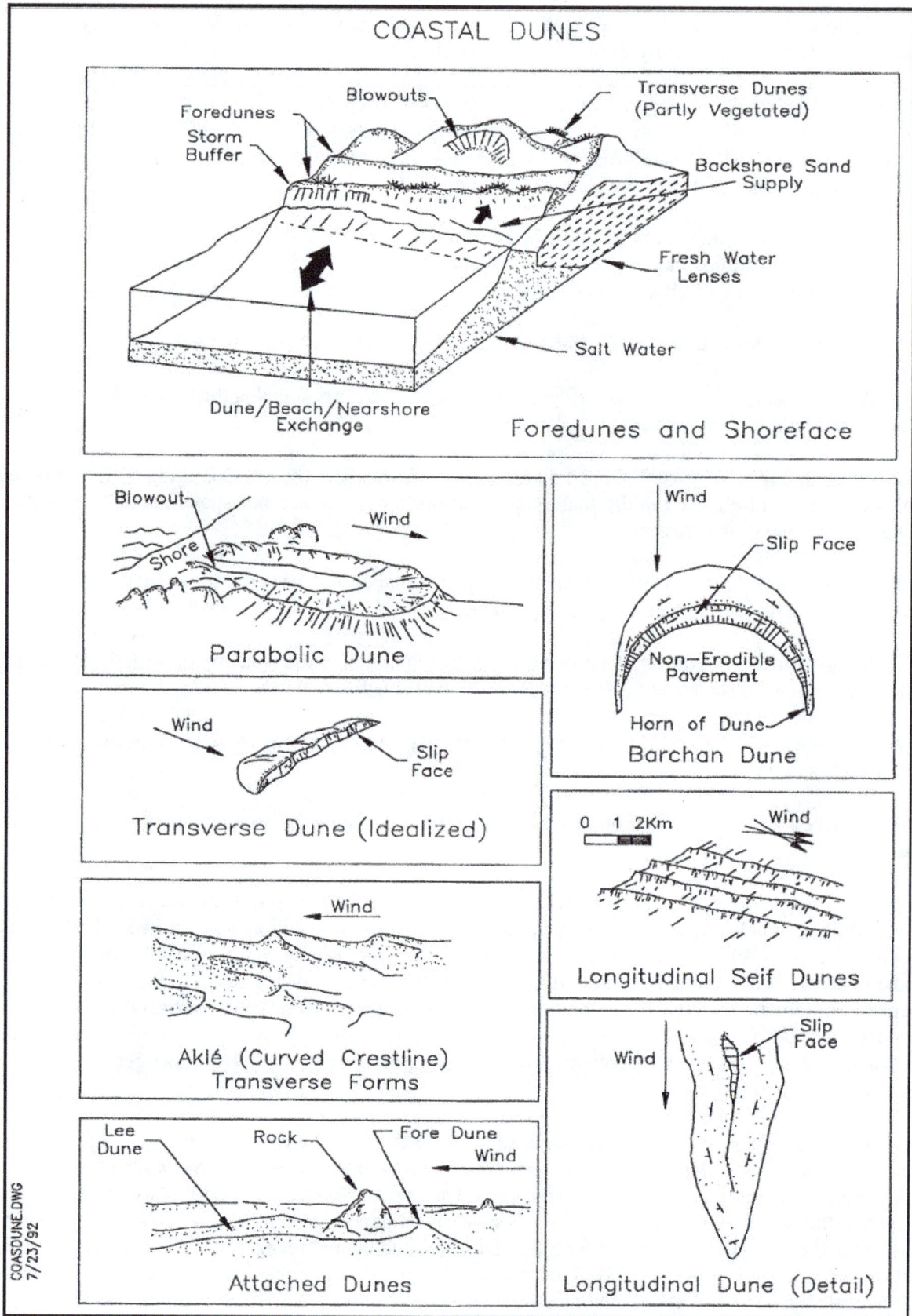

Figure IV-2-11. Variety of dune types (adapted from Carter (1988), Reading (1986), and Flint (1971))

Coastal Classification and Morphology

Figure IV-2-12. Blowout in dunes. Eastern Alabama on the Morgan peninsula east of the mouth of Mobile Bay (April 1995)

Figure IV-2-13. Huge dune anchored to a rock outcrop in the eastern Sinai Desert (January 1979). The dune is more than 15 m high. Most of valley floor is hard-packed sand, although there are pockets of loose sand that impede vehicle traffic. As of 1979, many of the bedouins in this area were already using pickup trucks instead of camels

Coastal dune management and conservation practices are reviewed in Carter, Curtis, and Sheehy-Skeffington (1992).

### IV-2-7. Volcanic Coasts

*a. Introduction and definitions.* *Volcanoes* are vents in the earth's surface through which magma and associated gases and ash erupt (Bates and Jackson 1984). Often, conical mountains are formed around the vents as repeated eruptions deposit layer upon layer of rock and ash. Therefore, the definition is extended to include the hill or mountain built up around the opening by the accumulation of rock materials.

(1) The fundamental importance of volcanism to mankind has been clearly documented around the world. The entire west coast of the Unites States is highly active tectonically and most of the continent's volcanoes are within 200 km of the coast. There are more than 260 distinct volcanoes younger than five million years in the Unites States and Canada, most of which are in Alaska and the Hawaiian Islands (Wood and Kienle 1990). Fifty-four have erupted in historic times, and distant memories of others are recounted in Native American legends.

(2) Volcanoes are important to coastal studies for many reasons:

(a) They provide sediment to the littoral environment. Material may reach the coast directly via ash fall-out and lava flows or may be transported by rivers from an inland source (e.g., Mount St. Helens).

(b) Vulcanism affects coastal tectonics (e.g., west coasts of North and South America).

(c) Shoreline geometry is affected by the formation of volcanic islands (Aleutians) and by lava that flows into the sea (Hawaiian Islands).

(d) Shoreline erodability ranges from very erodable for ash and unconsolidated pyroclastic rubble to very resistant for basalt.

(e) Volcanoes can pose a serious threat to coastal communities.

(f) Volcanic debris can choke rivers and harbors.

(3) This section briefly discusses general concepts of volcanism and describes features unique to volcanic shores. Examples from Alaska and the Hawaiian Islands illustrate the differences between composite and shield volcanoes and their associated coastlines. For the general reader, *Exploring our Living Planet*, published by the National Geographic Society (Ballard 1983), is a readable and interesting introduction to plate tectonics, hotspots, and volcanism.

*b. General geology.* Two classes of volcanoes can be identified, based on the explosiveness of their eruptions and composition of their lava. The ones in the Aleutians and along the west coasts of North and South America are known as *composite* volcanoes and are renowned for their violent eruptions. The paroxysmal explosion of Mount St. Helens on May 18, 1980, which triggered devastating mudflows and floods, killing 64 people, serves as a remarkable example. Composite eruptions produce large amounts of explosive gas and ash and build classic, high-pointed, conic mountains. In contrast, the Hawaiian Islands are *shield* volcanoes: broad, low, basalt masses of enormous volume. Shield eruptions are typically nonexplosive, and the highly liquid nature of their lava[1] accounts for the wide, low shape of the mountains. Volcanism affects the shore on two levels:

(1) The large-scale geologic setting of the continental margin affects sedimentation and overall coastal geology. Margins subject to active tectonism (and volcanism) are typically steep, with deep water occurring close to shore. Rocks are often young. High mountains close to shore provide a large supply of coarse sediments, and muddy shores are rare. Much sediment may be lost to deep water, particularly if it is funneled down submarine canyons. This is a one-way process, and the sediment is permanently lost to the coastal zone.

(2) Small-scale structures on volcanic shores may differ from those on clastic passive margins. Sediment supply may be frequently renewed from recent eruptions and may range greatly in size. Ash may be quickly destroyed in the sea, while basalt boulders may be tremendously resistant. Hardened shores at the sites of recent lava flows are difficult settings for harbor construction.

c.  *Composite volcanoes - coastal Alaska.* The coastal geology of Alaska is incredibly complex, having been shaped by fault tectonics, volcanism, glaciation, fluvial processes, sea level changes, and annual sea ice. More than 80 volcanoes have been named in the Aleutian arc, which extends for 2,500 km along the southern edge of the Bering Sea and the Alaskan mainland (Wood and Kienle 1990). Over 44 have erupted, some repeatedly, since 1741, when written records began. Aleutian arc volcanism is the result of the active subduction of the Pacific Plate beneath the North America Plate (Figure IV-2-14).

(1) Volcanoes have influenced the Aleutian Arc in two ways. First, they have been constructive agents, creating islands as eruption after eruption has vented rock and ash. In some places, fresh lava or mudflows accompanying eruptions have buried the existing coast, extending the shore seaward. The eruptions of Mts. Katmai and Novarupta in 1912 produced ash layers 3 to 15 m thick. The Katmai River and Soluka Creek carried vast amounts of loose ash to the sea, filling a narrow bay and burying a series of old beach ridges (Shepard and Wanless 1971). Usually, loose mudflow and ash deposits are reworked rapidly by waves, providing sediment for beach development. In addition, for years after an eruption, streams may carry rock and ash to the coast, allowing the coast to locally prograde. The second effect has been destructive, and small islands have been largely destroyed by volcanic explosions. Bogoslof, in the eastern Aleutians, is an example in which both rapid construction and destruction have influenced the island's shape over time (Shepard and Wanless 1971).

(2) Clearly, a history of volcanic instability would be a major consideration for a coastal engineer planning a harbor or project. Most new volcanic islands are uninhabited, but harbors may be needed for refuge, military, or commercial purposes. During World War II, air fields and harbors were built quickly on formerly uninhabited Aleutian Islands. Some islands can supply stone for construction at other locations, requiring loading facilities for boats or barges.

d.  *Shield volcanoes - Hawaii.* Each of the Hawaiian islands is made up of one or more huge shield volcanoes rising from the ocean floor. The islands are at the southern end of a chain of seamounts that extends 3,400 km to the northwest and then turns north and extends another 2,300 km toward Kamchatka as the Emperor Seamounts. More than 100 volcanoes, representing a volume of greater than one million cubic kilometers, make the Hawaiian-Emperor chain the most massive single source of volcanic eruption on earth (Wood and Kienle 1990). The submerged seamounts become successively older away from Hawaii.

---

[1] *Lava* is the term used for molten rock (and gasses within the liquid) that have erupted onto the earth's surface. *Magma* refers to molten rock that is still underground.

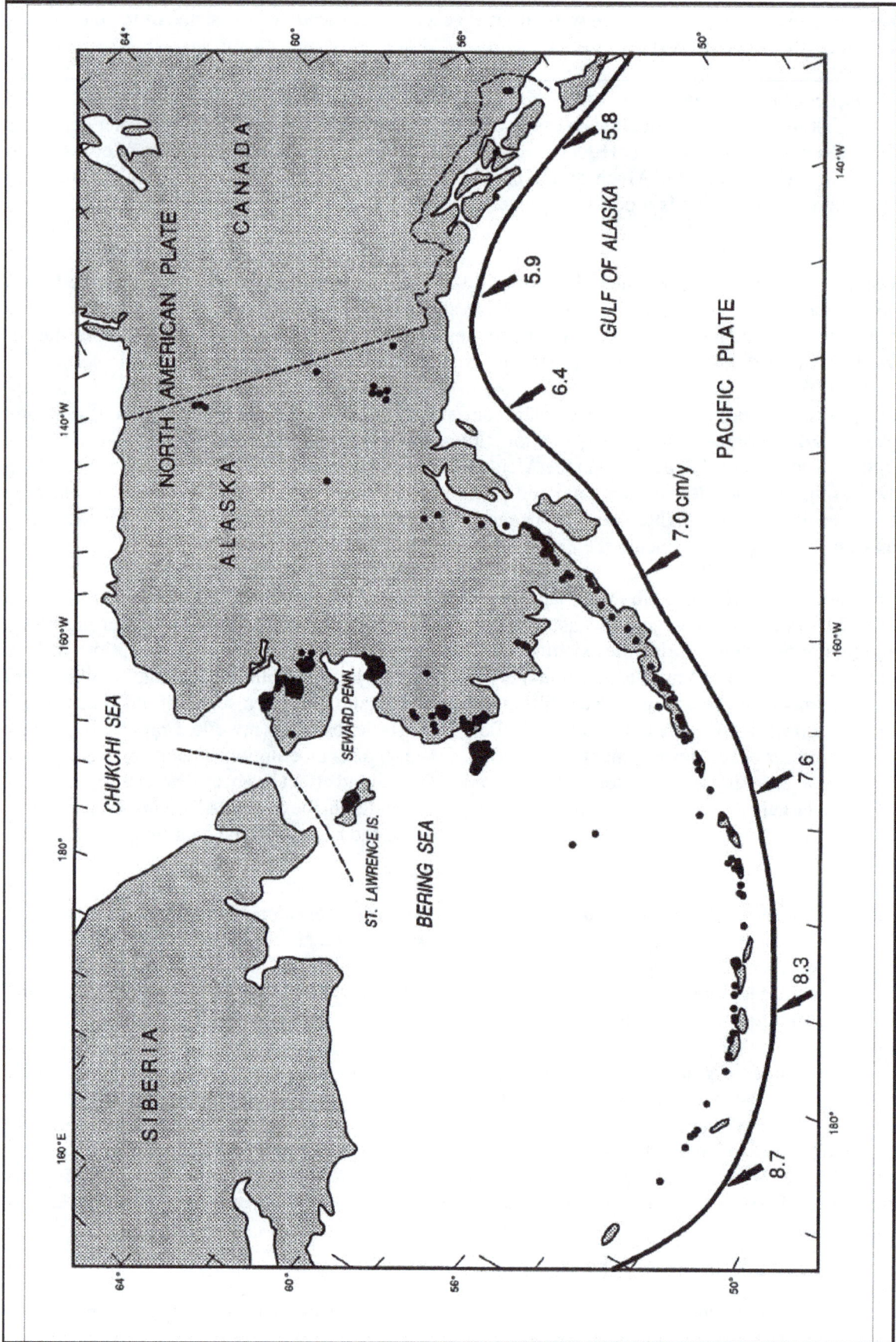

Figure IV-2-14. Alaskan volcanoes along the Aleutian Island arc, marking the boundary between the North American and the Pacific crustal plates. Arrows indicate subduction of the Pacific Plate in cm/year

Meiji Seamount, about to be subducted beneath Kamchatka, is 75-80 million years (my) old, Kilauea is only 0.4 my, while Loihi Seamount, south of the big island of Hawaii, is the newest member of the chain and has not yet emerged from the sea. The islands are found over a semipermanent "hot spot," a site where it is believed that a plume of hot, geochemically primitive material rises convectively through the mantle, interacts with the lithosphere, and vents on the seafloor (Dalrymple, Silver, and Jackson 1973). The Pacific plate is postulated to be moving over the hot spot at a rate of about 13 cm/yr, based on ages of the major vents on Hawaii (Moore and Clague 1992).

(1) Although the coastlines of the Hawaiian Islands are geologically young, wave erosion and the growth of coral reefs have modified most of the shores. Coastal plains have formed around the base of some volcanoes and between others (for example, the intermontaine plateau between Koolau and Waianae on Oahu). The plains are partly alluvial and partly raised reefs (Shepard and Wanless 1971). The greater parts of the Hawaiian coasts are sea cliffs, some as high as 1,000 m on the windward sides of the islands. There are also extensive beaches, the best of which tend to be on the western sides of the islands, protected from waves generated by the northeast trade winds. On southwestern Kauai near Kekaha, there are prograding beach ridges. Surprisingly, most of the beaches are composed not of volcanic debris but primarily of biogenic sediment. The rare volcanic sand beaches are found at the mouths of the larger rivers or along coasts where recent lava flows have killed the coral reefs (Shepard and Wanless 1971). Many beaches are undergoing serious erosion, and finding suitable sources of sand for renourishment has been difficult. This is a critical problem because tourism is a major part of the Hawaiian economy, and the beaches are among the great attractions.

(2) An example from the island of Hawaii helps illustrate the rugged nature of these volcanic shores. Hawaii, at the southeast end of the island chain, has been built up from at least seven independent volcanoes (Moore and Clague 1992). Mauna Loa, a huge dome at the southern end of the island, rises to 4,100 m above the sea (8,500 m above the seafloor). Kilauea, a low dome that rises out of the southeast side of Mauna Loa, has had a remarkable history of eruptions since 1800. Because of the porosity of the lavas, few permanent streams flow down the island, although rainfall on the windward side is heavy. The southeast coast of the island is a barren, rugged rock shore built up from hundreds of Kilauea lava flows (Figure IV-2-15). In Figure IV-2-15, the foreground is cracked, barren basalt, while the plateau in the background supports a cover of grass. The vertical cliffs are about 10 m high and in areas have been notched or undercut by the surf. Small steep pocket beaches consisting of black volcanic sands have developed in some notches. Because of the harsh wave climate and tectonic instability, coastal engineering in the Hawaiian Islands is particularly challenging (Figure IV-2-16).

e. *Hazards posed by volcanoes.*

(1) Introduction. Coastal projects and communities are subject to four general types of hazards caused by volcanic eruptions:

(a) Explosion-generated tsunamis that can flood coastal areas.

(b) Direct burial by lava or ash (recently experienced in Hawaii, Iceland, and Sicily).

(c) Burial or disruption by mudflows and fluvial sediment from inland eruptions, and changes in stream drainage and coastal sediment-discharge patterns.

(d) Loss of life and destruction from explosions.

**Figure IV-2-15. Southeast coast of Hawaii, near Kalpana. Rugged cliffs are built up of many lava flows. Small pocket beaches form between headlands when the cliffs are undermined and loose sediment accumulates**

Volcanoes seem a remote hazard to most people, but the danger is imminent and real to those who live in certain parts of the earth, especially along the boundaries of the earth's tectonic plates. Fortunately, fewer than 100 people have been killed by eruptions in Hawaii, where the volcanism is less explosive (Tilling, Heliker, and Wright 1987).

(2) Earthquakes and tsunamis.

(a) *Tsunamis* are waves created by ocean-bottom earthquakes, submarine landslides, and volcanic explosions. These long-period waves can travel across entire oceans at speeds exceeding 800 km/hr, causing extensive damage to coastal areas. The cataclysmic explosion of Krakatau on August 27, 1883, generated waves 30 m high that swept across the Sunda Strait, killing 36,000 coastal villagers on Java and Sumatra. One explosion produced one of the loudest noises in history, which was heard at a distance of 4800 km. The Hawaiian islands are particularly vulnerable to tsunamis caused by disturbances around the Pacific rim. The tsunami of April 1, 1946, generated towering walls of water that swept inland, damaging many coastal structures on the islands. In areas, the water rose to 16 m above the normal sea level. Photographs of the waves and the resulting damage are printed in Shepard and Wanless (1971) (Francis Shepard was living on Oahu at the time and vividly describes how the waves smashed his bungalow, forcing him and his wife to flee for their lives). On July 17, 1998, a wall of water 7 m high hit the northwest coast of Papua, New Guinea without warning, following a 7.0 magnitude earthquake about 30 km off the coast in the Pacific Ocean. Some 6,000 villagers perished in this tragedy.

Figure IV-2-16. Laupahoehoe Harbor, Island of Hawaii, Hawaii. Found on the steep and rugged northeast coast of Hawaii, this harbor provides the only access to the fertile Hamakua coast fishing grounds. This is a hostile environment, as shown by the basalt boulders tossed onto the boat ramp by storm waves

(b) Clearly, there is little that can be done to protect against the random and unpredictable tsunamis. A warning network has been established to notify people around the Pacific of earthquakes and the possibility that destructive waves may follow. Coastal residents are urged to heed these warnings!

(3) Ash and fluvial sediment. When Mount St. Helens exploded on May 18, 1980, 390 m of the top of the mountain was blown off, spewing a cloud of dust and ash high into the stratosphere. From its north flank, an avalanche of hot debris and scalding gasses created immense mudflows, burying the upper 24 km of the North Toutle valley to a depth of 50 m. Lahars, formed from dewatering of the debris avalanche, blocked the shipping channel of the Columbia River. This created an enormous dredging task for the U.S. Army Corps of Engineers, and ultimately much of the dredged material had to be disposed at sea. Dredging related to the explosion continues 18 years after the eruption, as material continues to move downstream from mountain watersheds.

(4) Explosive destruction. Communities close to volcanoes may be destroyed by the explosion and the inhabitants killed by poisonous gasses and superheated steam.

(a) The coastal example frequently cited is the destruction of St. Pierre on Martinique by the violent explosion of Montagne Pelée on May 8, 1902. A glowing cloud overran St. Pierre and spread fan-like over the harbor. Practically instantly, the population of more than 30,000 was smothered with toxic gas and incinerated (Bullard 1962).

(b) The cloud that destroyed St. Pierre consisted of superheated steam filled with even hotter dust particles, traveling more than 160 km/hr. The term *nuée ardente* is now used to describe this type of swiftly flowing, gaseous, dense, incandescent emulsion. It is also used as a synonym for the Peléan type of eruption.

### IV-2-8. Sea Cliffs - Diastrophic, Erosional, and Volcanic

*a.* Sea cliffs are the most spectacular geomorphic features found along the world's coastlines. This section concentrates on bedrock cliffs, with *bedrock* defined as "the solid rock that underlies gravel, soil, or other superficial material" (Bates and Jackson 1984). Bedrock cliffs are found along most of the U.S. and Canadian Pacific coast, in Hawaii, along the Great Lakes shores, and in Maine. South of Maine along the Atlantic coast, cliffs are rare except for some examples in New Hampshire, Massachusetts, and Rhode Island. The eastern end of Long Island, Montauk Point, consists of rapidly eroding till bluffs (Figure IV-2-9). Cliffs constitute portions of the coastlines of Spain, Italy, Greece, Turkey, Iceland, and the South American nations facing the Pacific Ocean. Shorelines with cliffs may be both emergent or submergent. For more information, Trenhaile's (1987) *The Geomorphology of Rock Coasts* presents a comprehensive and global review of cliffs, shore platforms, and erosion and weathering processes.

*b.* Bedrock cliffs are composed of all three major rock types: igneous, sedimentary, and metamorphic.

(1) **Intrusive igneous rock**, such as granite, cools and solidifies beneath the earth's surface, while *extrusive igneous rock*, such as basalt, is formed by lava above ground (it may erupt underwater or on land). Usually, igneous rocks are highly resistant; however, two properties affect their susceptibility to weathering and erosion (de Blij and Muller 1993):

(a) *Jointing* is the tendency of rocks to develop parallel sets of fractures without obvious external movement like faulting.

(b) *Exfoliation*, caused by the release of confining pressure, is a type of jointing which occurs in concentric shells around a rock mass.

(2) **Sedimentary rock** results from the deposition and lithification (compaction and cementation) of mineral grains derived from other rocks (de Blij and Muller 1993). This category also includes rock created by precipitation (usually limestone).

(a) The particles (clasts) that make up *clastic sedimentary rock* can range in size from windblown dust to waterborne cobbles and boulders. The vast majority of sedimentary rocks are clastic. Common examples include sandstone, composed of lithified sand (usually consisting mostly of quartz), and shale, made from compacted mud (clay minerals). Many cliffs along the south shore of Lake Erie are shale.

(b) *Nonclastic sedimentary rocks* are formed by precipitation of chemical elements from solution in marine and freshwater bodies because of evaporation and other physical and biological processes. The most common nonclastic rock is limestone, composed of calcium carbonate ($CaCO_3$) precipitated from seawater by marine organisms (and sometimes also incorporating marine shell fragments). Many of the Mediterranean cliffs are limestone and are very vulnerable to dissolution.

(3) **Metamorphic rocks** are preexisting rocks changed by heat and pressure during burial or by contact with hot rock masses. Common examples include:

(a) *Quartzite*, a very hard, weathering-resistant rock, formed from quartz grains and silica cement.

(b) *Marble*, a fine-grained, usually light-colored rock formed from limestone.

(c) *Slate*, a rock that breaks along parallel planes, metamorphosed from shale.

*c.* Sea cliffs are formed by three general processes:

(1) Volcanic eruptions and uplift caused by local volcanism (discussed in paragraph IV-2-7).

(2) Diastrophic activity that produces vertical movement of blocks of the crust.

(3) Erosional shorelines - partial drowning of steep slopes in hilly and mountainous terrain and resulting erosion and removal of sediment.

*d.* Sea cliffs, often found on tectonically active coasts, may be created by two mechanisms. First, if a block of the coast drops, a newly exposed fault plane may be exposed to the sea. The opposite process may occur: a block may be uplifted along a fault plane, exposing a formerly exposed portion of the shoreface to marine erosion. Older cliffs may be raised above sea level and be temporarily protected from further erosion. Earlier shorelines, sometimes tens of meters above the present sea level, are marked by notches or wave-cut platforms (sometimes termed uplifted marine terraces) (Figure IV-2-17). Terraces marking the highstand of eustatic (absolute) sea level have been traced around the world. Deep water is often found immediately offshore of faulted coasts. Cliffs that extend steeply into deep water are known as plunging cliffs.

*e. Erosional coasts* may be straight or may be irregular, with deeply indented bays. The way the shore reacts to inundation and subsequent marine erosion depends on both the wave climate and the rock type.

(1) Wave-straightened coasts. Cliffs are often found along shores where wave erosion rather than deposition is the dominant coastal process. Exposed bedrock, high relief, steep slopes, and deep water are typical features of erosional shorelines (de Blij and Muller 1993). When islands are present, they are likely to be remnants of the retreating coast rather than sandy accumulations being deposited in shallow water. The sequence of events that create a straightened coast are illustrated in Figure IV-2-18. The original coastline includes headlands and embayments (a). As waves attack the shore, the headlands are eroded, producing steep sea cliffs (b). The waves vigorously attack the portion of the cliff near sea level, where joints, fissures, and softer strata are especially vulnerable. The cliffs are undermined and caves are formed. Pocket beaches may accumulate between headlands from sediment carried by longshore currents. Especially durable pinnacles of rock may survive offshore as stacks or arches. Over time, the coast is straightened as the head-lands are eroded back (c).

(a) Beaches. Beaches may form at the base of cliffs if the rubble that has fallen from the cliff face (known as talus) is unconsolidated or friable and breaks down rapidly under wave attack (Figures IV-2-9 and IV-2-15). If the rock debris is durable, it may serve to armor the shore, protecting it from further wave attack except during the most severe storms.

(b) Wave-cut platforms. At the base of cliffs that have been progressively cut back by waves, near-horizontal platforms may form just below sea level. These rocky platforms may be of substantial width, depending on lithology and the time that sea level has been at that height (Figure IV-2-17). The platforms may be clean or may be covered with rubble fallen from the adjacent cliffs.

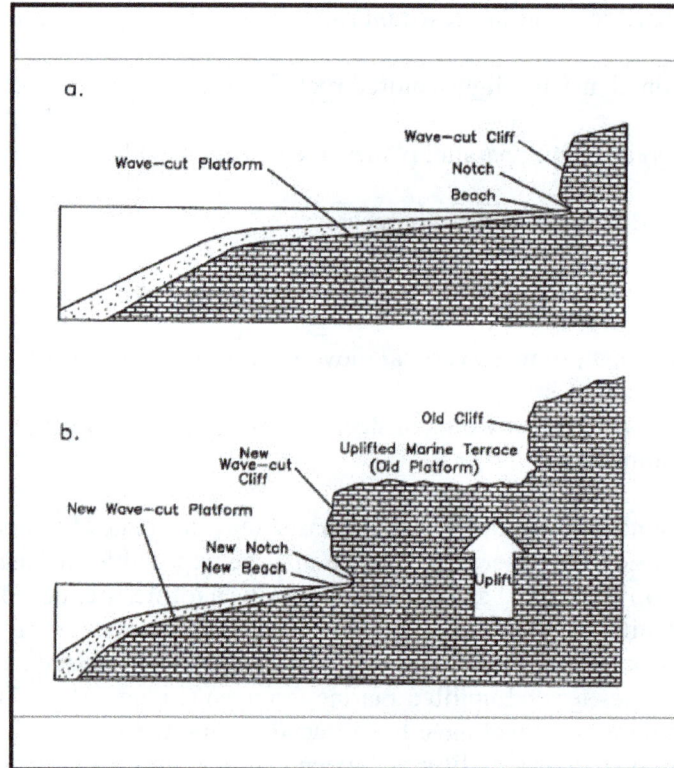

**Figure IV-2-17. Wave-cut platform exposed by tectonic uplift. Sediment that accumulated on the platform may temporarily protect the cliff from further erosion**

(2) Creation of irregular shorelines. In some mountainous terrains, rising sea level results in deeply incised coastlines. This process is illustrated in Figure IV-2-19. As the sea rises, a river valley is inundated. Once exposed to the sea, the new shoreline is subject to dissolution and biological attack. In southern France, Italy, Greece, and Turkey, thousands of deep embayments are found in the coastal limestone hills. The fact that the wave climate in the Mediterranean is relatively calm (compared with the open oceans) suggests that erosional processes other than wave attack have been instrumental in creating these steep, indented shores. An irregular shore may also be formed when differing rock types outcrop at the coast. Massive rocks, especially igneous and metamorphic ones, withstand erosion better than most sedimentary rocks, which are usually friable and contain bedding planes and fractures. The Pacific coast of Washington off the Olympic Peninsula is irregular because of the complex geology and variety of exposed rock formations (Figure IV-2-20).

*e.* Marine cliffs are degraded by many physical and biological factors.

(1) Wave attack is most likely the primary mechanism that causes cliffs to erode (Komar 1976). The hydraulic pressure exerted by wave impact reaches values sufficient to fracture the rock. Sand and rock fragments hurled at the cliff by waves grind away at the surface. Komar (1976) states that wave erosion occurs chiefly during storms, but admits that little quantitative research has been conducted. Once a cliff has been undercut at its base, the overlying rock, left unsupported, may collapse and slide down to the shoreline (Figures III-5-21, IV-2-21, and IV-2-22). Temporarily, the talus protects the cliff, but over time the rubble is reduced and carried away, leaving the fresh cliff face exposed to renewed wave attack.

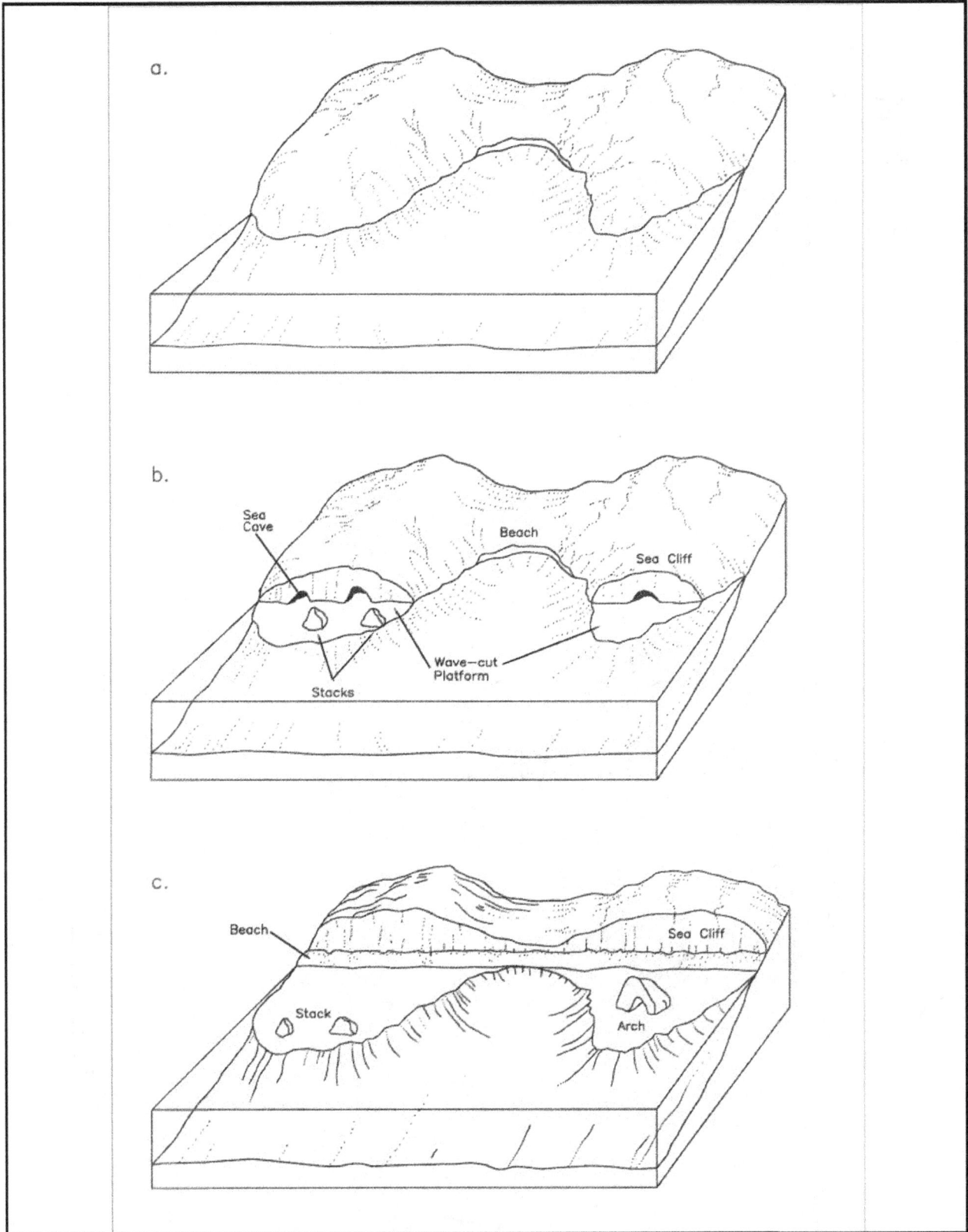

Figure IV-2-18. Wave erosion of an indented coastline produces a straightened, cliff-bound coast. Wave-cut platforms and isolated stacks and arches may remain offshore. Many such features are found along the Washington, Oregon, and California coast (adapted from de Blij and Muller (1993))

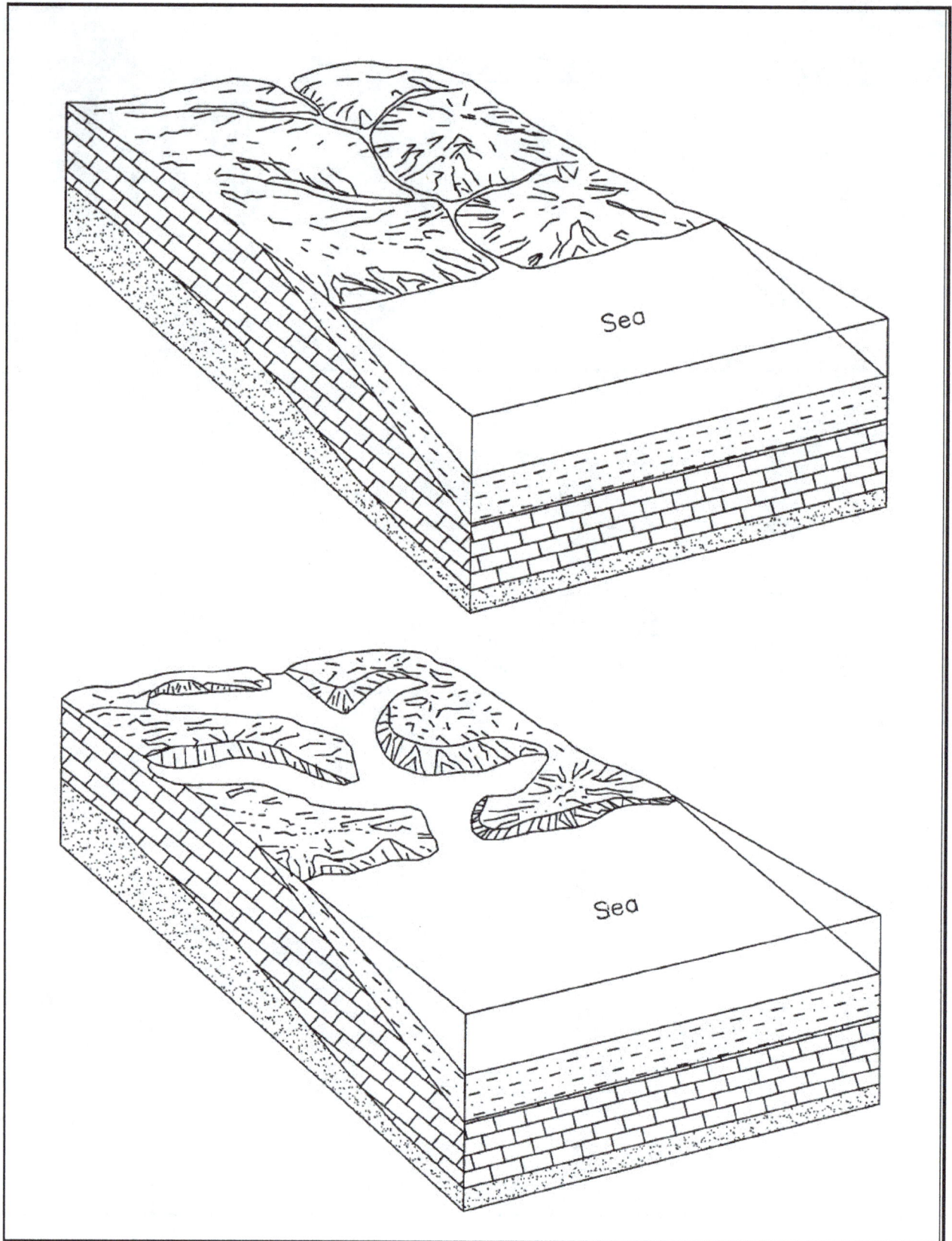

Figure IV-2-19. Inundation of a mountainous area by rising sea level or land subsidence produces a deeply indented shoreline

Figure IV-2-20.   Pacific Ocean coast of Olympic Peninsula, Washington, near Jefferson Cove (September 1971).   Rugged shore consists of resistant rock headlands with pocket beaches.   Much of the beach sediment is cobble and coarse clasts derived from the adjacent cliffs.   The beaches are often covered with logs and driftwood

(2)  In addition to waves, weathering processes weaken and crumble sea cliffs.  Ice wedging in cold climates progressively weakens the rock.  Plant roots grow and expand in cracks.  Lichens secrete acids that etch the rock surface.   Groundwater can lubricate impermeable rock surfaces, upon which large masses of overlying rock can slip.  This process is responsible for large slumps in the shale bluffs in Lakes Erie and Ontario (Figure III-5-21).

(3)  Mollusks and burrowing animals can weaken otherwise resistant massive rocks.  Komar (1976) lists burrowing mollusks such as *Pholadidae* and *Lithophaga*, and periwinkles, worms, barnacles, sponges, and sea urchins as having been observed to erode rock.  Boring algae can also weaken rock (Figure IV-2-23).

(4)  Under normal circumstances, surface seawater is saturated with calcium carbonate ($CaCO_3$), therefore minimizing dissolution of limestone or $CaCO_3$-cemented sediments.  Marine organisms can locally increase the acidity of the water in high-tide rock basins and other protected locations.  Small pockets found

Figure IV-2-21. A section of cliff, projecting out from the shore, is likely to collapse soon. To the left, rubble at sea level marks the location of a previous slump. The lower cliffs are poorly cemented conglomerate while the higher, vertical cliffs (above the trees), are limestone (near Nauplió, Greece, April 1992)

at water's edge, often housing periwinkles and other animals, may have been caused by biochemical leaching.

(5) Salt weathering is caused by the pressure exerted by NaCl and other salts in the capillaries of rocks. The weathering is caused by:

(a) Changes of volume induced by hydration.

(b) Expansion of salt crystals caused by temperature changes.

(c) Crystal growth from solution.

The main factor in determining the efficacy of chemical weathering is the amount of water available for chemical reactions and the removal of soluble products. This suggests, but does not necessarily restrict, that the greatest chemical weathering will occur in hot, humid climates (Trenhaile 1987).

## IV-2-9. Marine Deposition Coasts - Barriers

*a. Introduction.*

(1) Barriers are narrow, elongate sand ridges rising slightly above the high tide level and extending generally parallel with the coast, but separated from the mainland by a lagoon or marsh (Bates and Jackson

Figure IV-2-22. Lake Michigan shore south of St. Joseph, Michigan (November 1993). A triangle-shaped wedge of the bluff has recently slumped. The bluffs in this area have suffered rapid retreat, and many homes have been destroyed. At least three factors account for the erosion: (a) offshore downcutting of the till lake bottom; (b) wave attack on the bluffs; (c) ice expansion and lubrication of bedding planes caused by groundwater

1984). The term *barrier* identifies the sand ridges as ones that protect parts of the coast from direct wave attack of the open ocean. In this manual, barrier will refer to the overall structure (sometimes called a barrier complex), which includes the beach, submerged nearshore features, underlying sediments, and the lagoon between the barrier and the mainland (Figure IV-2-24). Inlets and channels can also be considered part of a barrier system.

(2) The term *beach* is sometimes used as a synonym for barrier, but this can lead to confusion because a beach is a geomorphic shore type found throughout the world, even on volcanic or coralline coastlines where barriers are rare. Whereas all barriers include beaches, not all beaches are barriers.

(3) The following sections describe general barrier island morphology, history, and formation, subjects that have fascinated geologists for more than 100 years. The emphasis will be on long-term changes, covering periods of years or centuries. The purpose is to explain factors that lead to barrier migration or evolution. Longshore sediment transport, details on the morphology of sandy shorefaces, and the normal effects of waves and tides will be covered in Part IV-3, "Coastal Morphodynamics." This distinction is arbitrary because, clearly, the day-to-day processes that affect beaches also influence barrier development. In addition, the evolution of barriers during the Holocene Epoch is intimately related to sea level changes (see Part IV-1). These factors underscore the complex interrelationships that exist throughout the coastal zone and the difficulty of separating the constituent elements.

Figure IV-2-23.  Cemented conglomerate with many pits and cavities shows evidence of dissolution.  The rock mass has been undercut over 1 m (near Nauplió, Greece).  Note that the conglomerate is distinctly graded, with fine grains near the bottom and cobble near the top

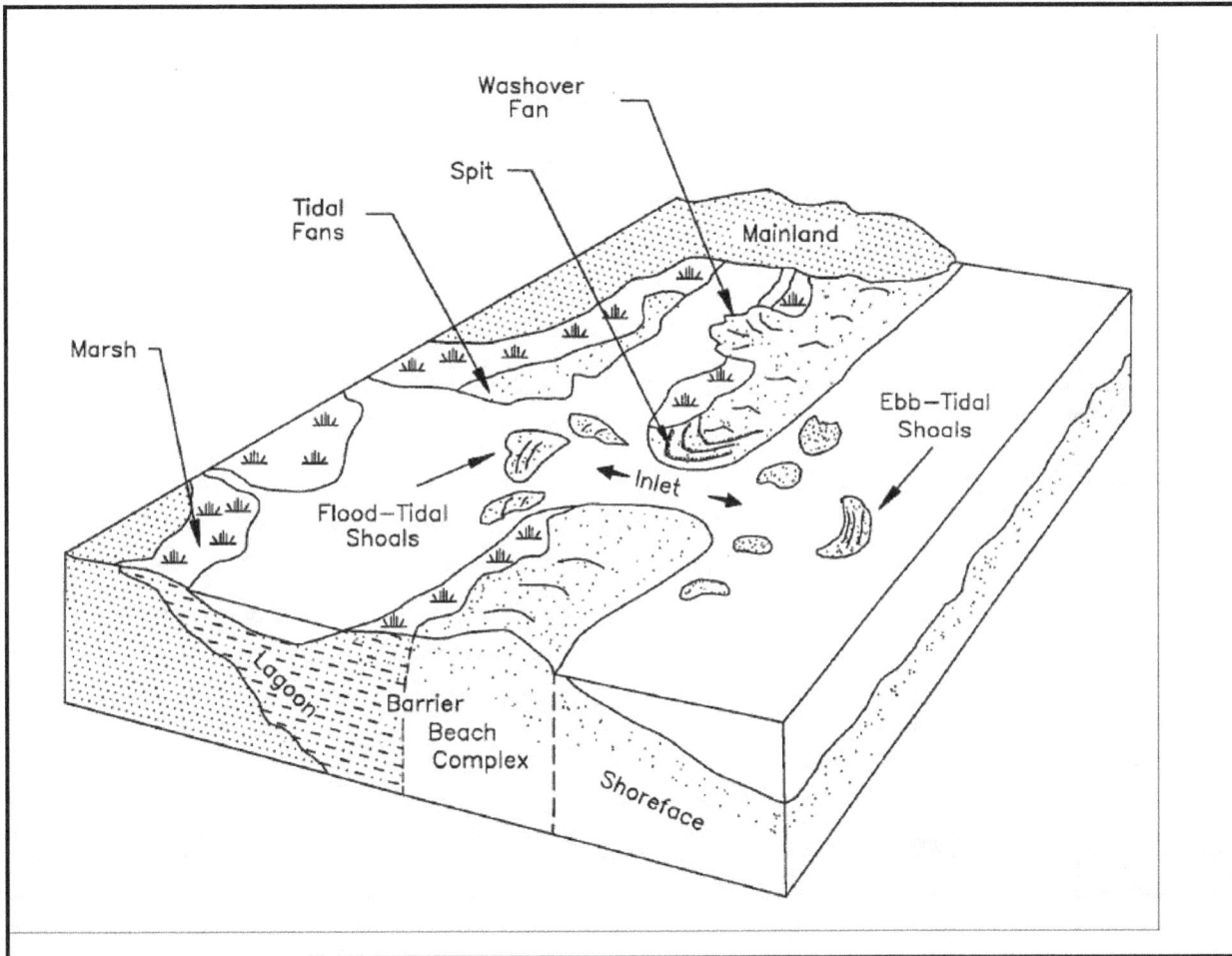

**Figure IV-2-24.** A three-dimensional view of features commonly found in barrier island systems, including the back barrier, overwash fans, and lagoons. The lagoons often become estuaries, particularly if sediment is supplied by both rivers and inlets

(4) The long-term and widespread interest in barrier islands is largely due to their great economic importance. Ancient buried barriers are important petroleum reservoirs. Contemporary barriers protect lagoons and estuaries, which are the breeding grounds for many marine species and birds. In addition, barrier islands are among the most important recreational and residential regions. In recent years, man's adverse impact on these fragile ecological and geological environments has led to increased need to study their origins and development to improve coastal management and preserve these critical resources for the future.

(5) An enormous literature on barrier islands exists. Nummedal (1983) provides a readable and concise overview. Leatherman's (1979) book is a compilation of papers on U.S. east coast and Gulf of Mexico barriers. Many seminal papers on barrier island evolution have been reprinted in Schwartz (1973). Textbooks by Carter (1988), Davis (1985), King (1972a), and Komar (1976) discuss barriers and include voluminous reference lists. Classic papers on beach processes have been reprinted in Fisher and Dolan (1977).

*b. Distribution of barrier coasts.* Barrier islands are found around the world (Table IV-2-2). They are most common on the trailing edges of the migrating continental plates (Inman and Nordstrom 1971)[1]. This type of plate boundary is usually non-mountainous, with wide continental shelves and coastal plains. Over 17 percent of the North American coastline is barrier, most of it along the eastern seaboard facing the Atlantic Ocean and the Gulf of Mexico. Table IV-2-3 lists the lengths of barriers and spits in the United States. Of the Atlantic states, Maine and New Hampshire have the fewest barriers because their coasts are largely composed of igneous rock. Massachusetts, with mostly glacial moraines and outwash along the coast, has the surprising total of 184 km of spits and barriers. Of the continental states, Florida has the most barriers and spits, totaling over 1,000 km for both the Atlantic and the Gulf of Mexico. Most of the shorelines facing the Gulf of Mexico consist of barrier islands. A portion of Florida's west coast, where wave energy is low, is mangrove swamp, but the Panhandle is famous for its glistening white barriers.

**Table IV-2-2**
**Worldwide Distribution of Barrier Island Coasts**

| Continent | Barrier Length (km) | % of World Total Barriers | % of Continent's Coastline that is Barrier |
|-----------|--------------------|--------------------------|---------------------------------------------|
| N. America | 10,765 | 33.6 | 17.6 |
| Europe | 2,693 | 8.4 | 5.3 |
| S. America | 3,302 | 10.3 | 12.2 |
| Africa | 5,984 | 18.7 | 17.9 |
| Australia | 2,168 | 6.8 | 11.4 |
| Asia | 7,126 | 22.2 | 13.8 |
| Total | 32,038 | 100.0 | |

From Cromwell (1971)

Almost the entire shore of Texas consists of long barriers, which continue south into Mexico. Extensive barriers are also found on the Gulf of Alaska north of Bering Strait. Including numerous spits in the Aleutians and the Gulf of Alaska, the state of Alaska has almost 1,300 km of barrier in total, exceeding Florida. The United States total shown in Table IV-2-3 is 4,882 km, about half the North American total computed by Cromwell (1971). For more information, the most extensive survey of United States barriers was documented in the *Report to Congress: Coastal Barrier Resources System* (Coastal Barriers Study Group 1988).

*c. General coastal barrier structure.* The barrier shore type covers a broad range of sizes and variations. Three general classes of barrier structures can be identified (Figure IV-2-25):

(1) Bay barriers - connected to headlands at both ends and enclosing a bay or wetland.

(2) Spits - attached to a sediment source and growing downdrift. May be converted to a barrier island if a storm cuts an inlet across the spit. May evolve into bay barriers if they attach to another headland and completely enclose a lagoon.

---

[1] *The trailing edge* of a continent is moving away from an active spreading center. For example, the Atlantic coast of the United States is a trailing edge because new seafloor is being formed along the mid-Atlantic ridge, causing the Atlantic Ocean to grow wider (Figure IV-1-2). The Pacific coast is a *leading edge* because the oceanic plates are being subducted (consumed) at various trenches and are therefore becoming smaller.

Coastal Classification and Morphology

Table IV-2-3
Barrier Islands and Spits of the United States

| Ocean or Sea | State | Total Length (km)[1] |
|---|---|---|
| Atlantic | Maine | 11.4 |
| | New Hampshire | 2.5 |
| | Massachusetts [2] | 184.4 |
| | Rhode Island [3] | 17.6 |
| | New York [4] | 152.2 |
| | New Jersey | 106.0 |
| | Delaware [5] | 33.7 |
| | Maryland [5] | 49.2 |
| | Virginia [5] | 126.0 |
| | North Carolina | 380.7 |
| | South Carolina | 234.2 |
| | Georgia | 159.0 |
| | Florida | 533.3 |
| | **Atlantic coast total** | **1990** |
| Gulf of Mexico | Florida | 478.5 |
| | Alabama | 92.7 |
| | Mississippi | 59.5 |
| | Louisiana | 151.9 |
| | Texas | 498.0 |
| | **Gulf of Mexico total** | **1281** |
| Pacific - Continental USA | Washington [6] | 63.9 |
| | Oregon | 91.9 |
| | California | 65.4 |
| | **Pacific total** | **221** |
| Beaufort, Chukchi, Bering Seas, Gulf of Alaska, Bristol Bay | Alaska total (incl. Aleutians) | **1266** |
| Lakes Superior, Huron, Michigan, Ontario, Erie | Combined Great Lakes states | **124** |
| **United States total** [2,3,4,5,6] | | **4882** |

Source: Unpublished data generated during the Corps of Engineers' Barrier Island Sediment Study (BISS), 1989.
[1] Length of barriers measured from U.S. Geological Survey topographic maps. Includes barriers and spits enclosing a body of water or marsh, not the total length of beaches in the United States. No data available for Puerto Rico, Virgin Islands, Pacific Trust Territories.
[2] Includes Nantucket and Martha's Vineyard Islands.
[3] Does not include Narragansett Bay.
[4] Atlantic Ocean only; does not include spits in Long Island Sound or Great Peconic Bay.
[5] Does not include Chesapeake Bay.
[6] Includes spits in Strait of Juan de Fuca. Does not include Long Beach Peninsula, enclosing Willapa Bay.

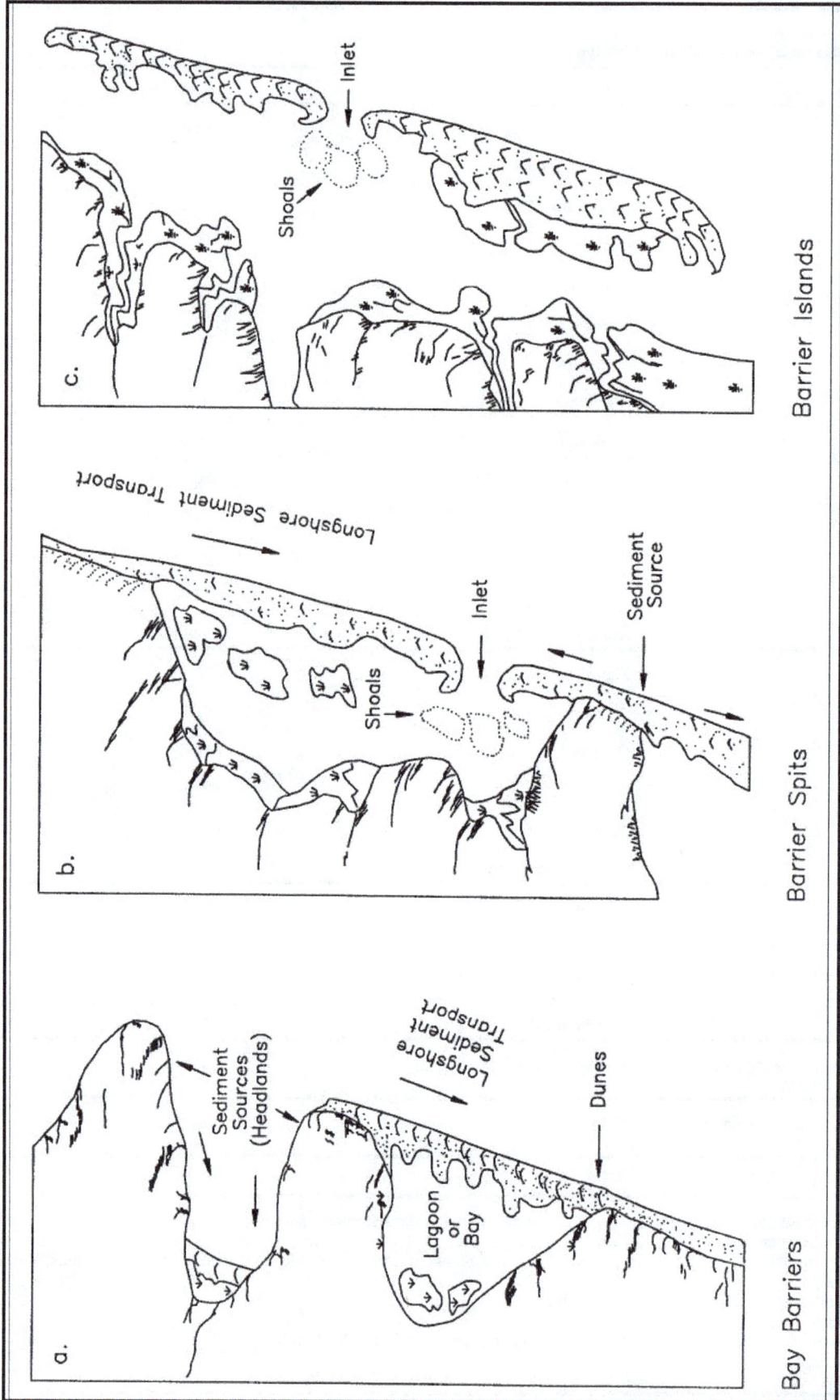

Figure IV-2-25. General barrier types: bay, spit, and island

(3) Barrier islands - linear islands that are not attached to the mainland but that enclose a bay, pond, or marsh/wetland. A series of these islands extending along the coast are defined as a barrier chain.

*d. Origin and evolution.* The origin of barrier islands has been a topic of debate among geologists for over a century (Schwartz 1973). The differing theories suggest that there are probably several types of barriers, each undergoing its own form of development due to unique physical and geologic factors. Three main theories have evolved, all of which have fierce supporters and critics.

(1) Emergence model. De Beaumont in 1845 was the first naturalist formally to present a theory of barrier island formation. It was supported and modified by the influential Johnson (1919). These researchers theorized that barrier emergence began with the formation of an offshore sand shoal, which consisted of material reworked from the seafloor by waves. Over time, the shoal would accumulate ever more sand and grow vertically, eventually emerging above the sea surface (Figure IV-2-26). Wave swash and wind deposition would continue to contribute sand to the shoal, allowing it to grow larger and larger. Hoyt (1967) objected to this hypothesis because he was unaware of any examples of bars emerging above water and surviving wave action, although the growth of submerged bars was well-recorded. Otvos (1970) reported evidence from the Gulf coast supporting the emergence of submarine shoals (he conveniently noted that subsequent migration of barriers might completely obscure the conditions of formation of the original barrier).

(2) Submergence model. The submergence concept was refined by Hoyt (1967) and has received much support. In this model, the initial physical setting is a mainland beach-and-dune complex with a marsh separating the beach from higher terrain inland. Rising sea level floods the marsh, creating a lagoon that separates the beach from the mainland (Figure IV-2-27). Presumably, usually the sea level rise is part of a worldwide pattern (eustatic), but it may be caused in part by local submergence. Once formed, maintenance of the barrier becomes a balance of sediment supply, rate of submergence, and hydrodynamic factors.

(3) Spit detachment model. The third major model calls for the growth of sandspits due to erosion of headlands and longshore sediment transport (Figure IV-2-28). Periodically, the spit may be breached during storms. The furthest portion of the spit then becomes a detached barrier island, separated by a tidal inlet from the portion that is still attached to the mainland. Gilbert (1885) may have been the first geologist to suggest the spit hypothesis, based on his studies of ancient Lake Bonneville, but the hypothesis lay dormant for many years because of Johnson's (1919) objections. In recent years, it has received renewed support because the cycle of spit growth and breaching can be seen in many locations (for example, at Cape Cod, Massachusetts (Giese 1988)).

(4) Combined origin model. Schwartz (1971) concluded that barrier island formation is most probably a combination of all of the above mechanisms. He felt that only a few examples of barriers could be cited as having been formed by only one method. Most systems were much more complex, as demonstrated by the barriers of southern Louisiana, which were formed by a combination of submergence and spit detachment (Penland and Boyd 1981).

*e. Barrier response to rising sea level.*

(1) Many of the barriers in the United States, particularly along the Atlantic coast, are eroding, causing serious economic and management challenges. What is responsible for this erosion?

(2) Sea level and sediment availability are probably the major factors that determine barrier evolution (Carter 1988). Three sea level conditions are possible: rising, falling, and stationary. Rising and falling sea

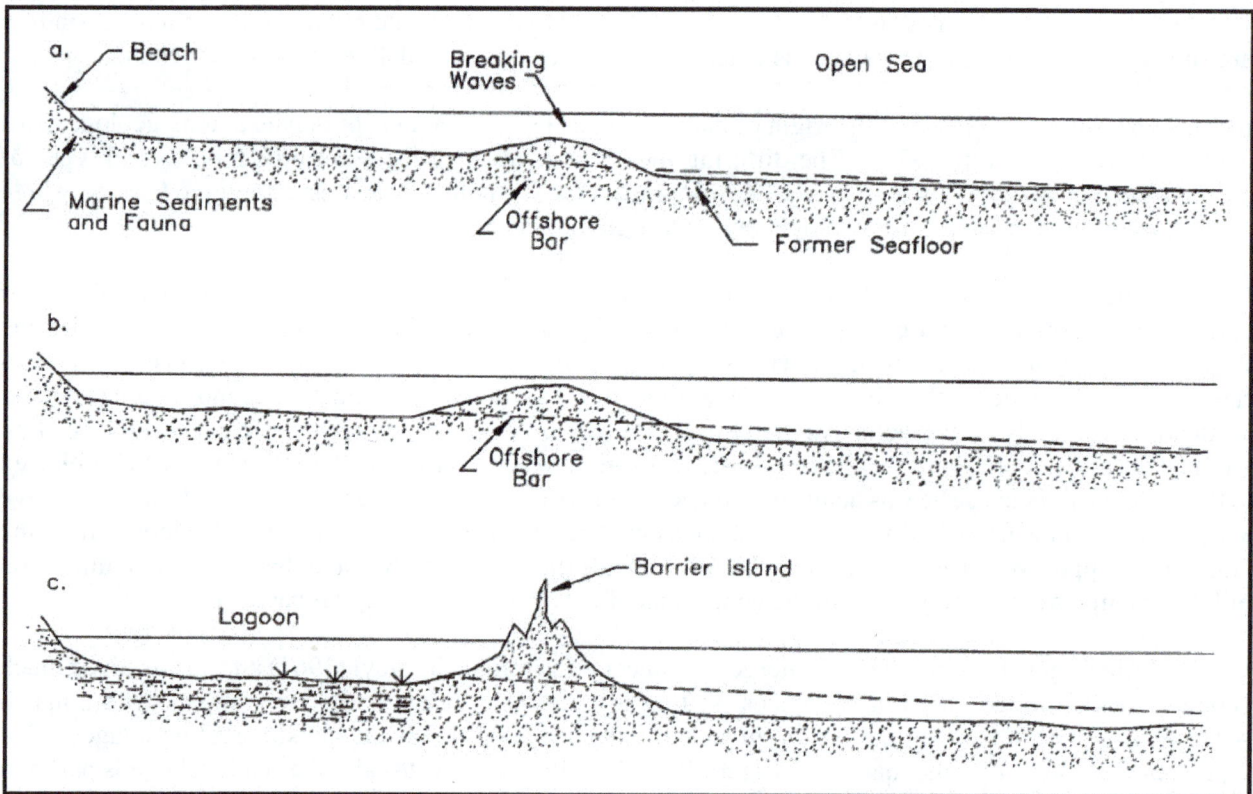

**Figure IV-2-26. Emergence model of barrier island formation (modified from Hoyt (1967)), (a) Waves erode the seafloor, forming a sandbar, (b) The bar continues to grow higher and wider, (c) Bar is converted to an island, enclosing a lagoon on the landward side**

usually result in great sediment movement as barriers adjust. A stationary stage, however, allows the shore to adjust slowly and achieve equilibrium between sediment supply and dynamic processes. Commonly, if sea level rises and sediment supply is constant, a barrier retreats (*transgression* of the sea). On the other hand, if sea level is rising but a large amount of sediment is supplied locally by rivers or eroding headlands, a particular barrier may be stable or may even aggrade upwards (see Table IV-1-6). However, many other factors can intervene: local geological conditions, biological activity, susceptibility to erosion, the rate of sea level change. Therefore, each location must be evaluated individually.

(3) Given the condition of rising sea level along the eastern United States, what are the mechanisms that cause barrier retreat? Three models of shoreline response to rising sea level have been proposed (Figure IV-2-29). These assume that an equilibrium profile is maintained as the shoreline is displaced landward and upward. In addition, overall sediment budget is balanced and energy input is constant.

(a) The first model, often called the Bruun Rule (Bruun 1962), assumes that sediment eroded from the shoreface is dispersed offshore. As water level rises, waves erode the upper beach, causing the shoreline to recede. Conceptually, this erosion supplies sediment for upward building for the outer part of the profile. The model assumes that the initial profile shape will be reestablished farther inland but at a height above the original position equal to the rise in water level $z$. Therefore, the retreat of the profile $x$ can be calculated from the following relationship (a modified version of the Bruun Rule):

Coastal Classification and Morphology

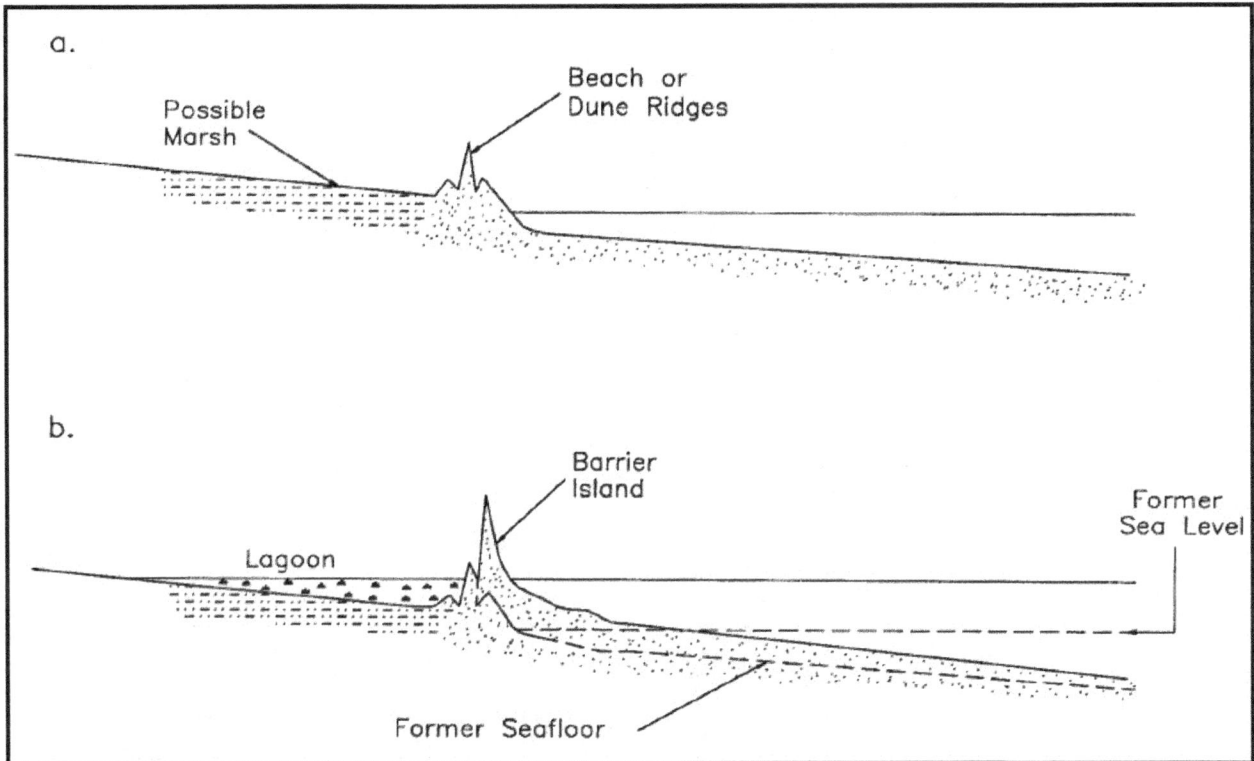

Figure IV-2-27. Submergence model of barrier island formation (modified from Hoyt (1967)), (a) Beach or sand dune ridges form near the shoreline, (b) Rising sea level floods the area landward of the ridge, forming a barrier island and lagoon. Sediment must be available to allow the island to grow vertically as sea level rises

$$x = \frac{zX}{Z}$$

(IV-2-1)

where the terms $x$, $z$, $X$, and $Z$ are shown in Figure IV-2-29a. Attempts to verify the Bruun rule have been ambiguous, and modifications to the model have been proposed (Dolan and Hayden 1983). The most successful studies required long-term data sets, such as the profiles from Lake Michigan examined by Hands (1983). His research showed that the shoreface profile requires a considerable time (years or decades) to adjust to water level changes. It is unclear whether the Bruun Rule would apply if an ample supply of sediment were available during rising sea level. Would the barrier essentially remain in place while sand eroded from the shoreface or newly supplied sand was dispersed offshore to maintain the profile? The Bruun Rule and some of its underlying assumptions are discussed in greater detail in Part IV-3.

(b) Landward migration of a barrier by the rollover model applies to coasts where washover processes are important. As sea level rises, material is progressively stripped from the beach and shoreface and carried over the barrier crest by waves. The sand is deposited in the lagoon or marsh behind the barrier (Figure IV-2-30). Dillon (1970) documented this process along the southern Rhode Island coast. As the barrier moves landward (rolls over itself), lagoonal sediments may eventually be exposed on open shoreface. Evidence of this can be seen in Rhode Island during winter storms, when large pieces of peat are thrown up on the beach. Dingler, Reiss, and Plant (1993) have described a model of beach erosion and overwash deposition on the Isles Dernieres, off southern Louisiana. They attributed a net annual beach retreat of

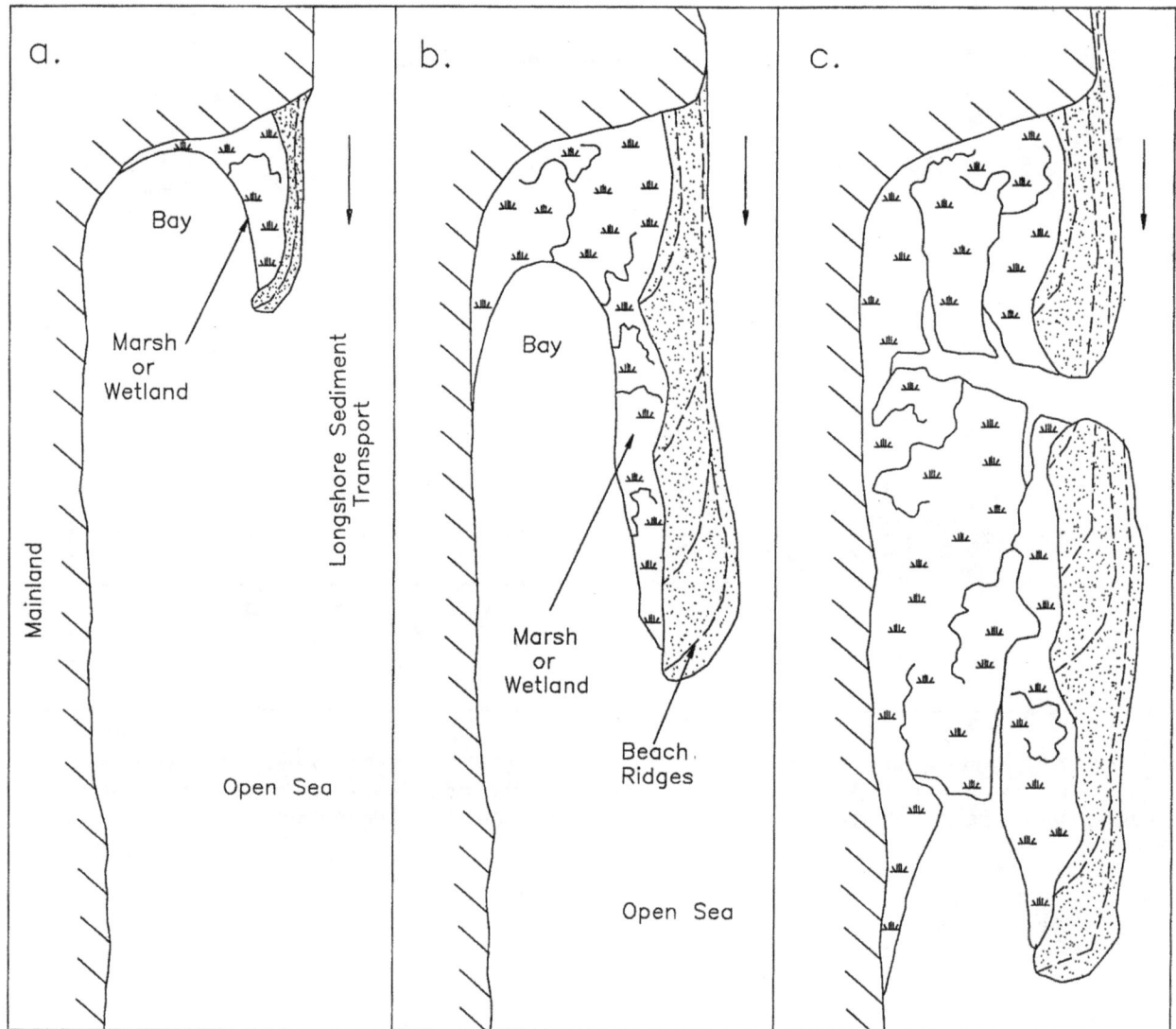

**Figure IV-2-28. Barrier island formation from spit (modified from Hoyt (1967)), (a) Spit grows in direction of longshore drift, supplied from a headland, (b) Spit continues to grow downdrift, marsh begins to fill semi-protected bay, (c) Part of spit is breached, converting it to a barrier island**

greater than 10 m/yr to winter cold-front-driven storms that removed sediment from the beach face and infrequent hurricanes that shifted a substantial quantity of sediment to the backshore. For the most part, rollover is a one-way process because little of the sand carried over the barrier into the lagoon is returned to the open shoreface.

(c) The barrier overstepping model suggests that a barrier may be drowned, remaining in place as sea level rises above it. Several hypotheses have been proposed to explain how this process might occur:

- If the rate of sea level rise accelerates, the barrier may be unable to respond quickly by means of rollover or other mechanisms. Carter (1988) cites research that suggests that gravel or boulder barriers are the most likely to be stranded.

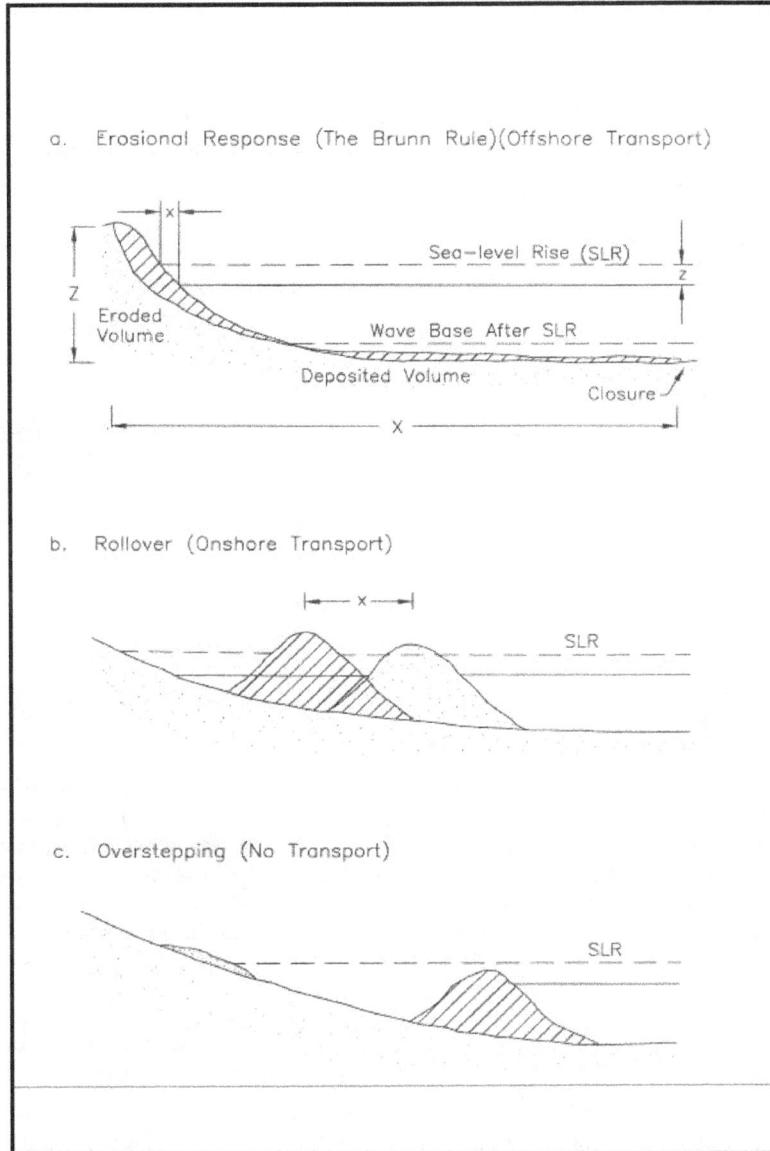

a. Erosional Response (The Brunn Rule)(Offshore Transport)

Sea-level Rise (SLR)

Eroded Volume

Wave Base After SLR

Deposited Volume

Closure

b. Rollover (Onshore Transport)

SLR

c. Overstepping (No Transport)

SLR

Figure IV-2-29. Three models of shoreline response to sea level rise: (a) Erosional response model/Bruun Rule assumes offshore dispersal of eroded shoreline materials; (b) Island rollover model assumes that barriers migrate landward according to the rate of sea level rise; (c) Overstepping model assumes submergence in place (adapted from Carter (1988))

- A modest influx of sediment may retard barrier migration enough to allow overstepping. If a constant volume of sediment is available, the new material must be distributed over a wider and wider base as sea level rises. The result is that vertical growth per unit time decreases. Eventually, the barrier is overtopped and the surf zone jumps forward to the bay shoreline that was formerly protected.

Figure IV-2-30. Westhampton Beach, Long Island, New York (March 1994 - view looking north towards Moriches Bay). In December 1992, the barrier island breached here during a northeaster, and 60 houses were destroyed. A large overwash fan can be seen in the bay, and to the left are houses that survived the storm. This is an example of the rollover mechanism, when sand from the ocean-facing shoreface is carried over the barrier and deposited in the back bay. The breach was repaired by the Corps of Engineers during 1993 using sand dredged from offshore. This low stretch of the barrier is particularly vulnerable, with several earlier breaches on record. In summer 1998, new homes were again being built here

- A barrier may remain in place because of a dynamic equilibrium that develops between landward and seaward sediment transport. As sea level rises, tidal prism of the lagoon increases, resulting in more efficient ebb transport. During this time, an increasing amount of washover occurs, but the effect is counteracted because sediment is being returned to the exposed shoreface. If little or no new sediment is added to the system, the sea eventually rises above the barrier crest, allowing the surf zone to jump landward to a new location (the formerly protected mainland shore).

- All three of these mechanisms may come into play at various times, depending upon environmental conditions. Sediment supply may be the crucial factor, however. Some stranded barriers, such as the ones in the northeastern Gulf of Mexico, may have maintained vertical growth because of an adequate sediment supply (Otvos 1981).

(4) In all likelihood, barriers respond to all three of the migration models, depending upon timing and local conditions such as sediment supply or preexisting topography (Carter 1988). During the initial stages of sea level rise, the shore erodes and material is dispersed offshore (the Bruun Rule). As the barrier becomes narrower, washover carries more and more sediment to the back lagoon. Eventually, the barrier may become stranded and be drowned. These models have been criticized because they are two-dimensional and do not account for variations in longshore drift. The criticisms are valid because drift is sure to vary greatly as barriers are progressively reshaped or drowned. As a result, pockets of temporarily prograding barriers may remain along a generally retreating coastline.

(5) In summary, several models have been advanced to explain how barrier islands respond to rising sea level. However, because the interactions in the coastal zone are so complex, trying to reduce barrier evolution to a series of simple scenarios is unrealistic. Much more research is needed to define the many factors that contribute to barrier evolution.

## IV-2-10. Marine Deposition Coasts - Beaches

*a. Introduction.* Marine and lacustrine beaches comprise one of the most widely distributed coastal geomorphic forms around the world. Their importance as a buffer zone between land and sea and as a recreational and economic resource has stimulated studies by earth scientists for well over a century. Although much has been learned about how beaches form and how they are modified, the coastal environment is incredibly complex, and each location responds to unique geologic conditions and physical processes. Some of these variable factors include:

(1) Seasonal cycles.

(2) Long-term trends.

(3) Changes in relative sea level.

(4) Variations in sediment supply.

(5) Meteorological cycles.

As a result, it is difficult to characterize beaches and predict future developments without the benefit of long-term studies and observations. The following sections describe the morphology and sediments of beaches and define terms (also see Table IV-1-1). For additional information and extensive bibliographies, the reader is referred to Carter (1988), Davis (1985), Komar (1976, 1983), and Schwartz (1973, 1982).

*b. General definition.* **Beach** is defined as a gently sloping accumulation of unconsolidated sediment at the edge of a sea or other large body of water (including lakes and rivers). The landward limit may be marked by an abrupt change in slope where the beach meets another geomorphic feature such as a cliff or dune. Although this landward boundary has been consistently accepted in the literature, the seaward limit has been more broadly interpreted. Some authors have included the surf zone and the bar and trough topography in their definition because the processes that occur in the surf zone directly affect the exposed portion of the beach. The length of beaches varies greatly. Some stretch for hundreds of kilometers, such as those on the Carolina Outer Banks. Others, called pocket beaches, are restricted by headlands and may be only a few tens of meters long.

*c. Major subdivisions.* Beaches are part of the littoral zone, the dynamic interface between the ocean and the land. The littoral zone is bounded on one side by the landward limit of the beach and extends tens or hundreds of meters seaward to beyond the zone of wave breaking (EM 1110-2-1502). Beaches can be divided into two major zones: the foreshore and the backshore.

(1) Foreshore.

(a) The foreshore extends from the low-water line to the limit of wave uprush at high water (Figure IV-1-2). The upper portion of the foreshore is a steep slope where the high water uprush occurs. The seaward, lower, portion of the foreshore is sometimes called the *low-water or low-tide terrace.* This terrace often features low, broad ridges separated by shallow troughs, known as ridges and runnels (Figure IV-2-31).

**Figure IV-2-31. Ridge and runnel system, low water terrace, Charlestown Beach, Rhode Island. Sea grass debris in the foreground marks the height of the wave runup during the previous high tide. Where should the "shoreline" be defined on this beach?**

Because the foreshore is frequently subject to wave swash, it usually has a smoother surface than the backshore. There may be a minor step near the low-water mark, called the *plunge step.* Often, shell or gravel is concentrated at the base of this step, while the sediments to either side are much finer.

(b) The foreshore is sometimes called the *beachface.* However, beachface is also used in a more restricted sense to designate the steepened portion of the upper foreshore where the high-water wave uprush occurs. Therefore, it is recommended that foreshore and beachface not be used synonymously and that beachface be restricted to its upper foreshore definition.

(2) Backshore.

(a) The backshore extends from the limit of high water uprush to the normal landward limit of storm wave effects, usually marked by a foredune, cliff, structure, or seaward extent of permanent vegetation. The backshore is not affected by waves regularly, but only during storms, when high waves and storm surges allow reworking of backshore sediments. Between inundations, the backshore develops a rough surface because of vehicle or animal traffic and the development of wind-blown bed forms. On eroding beaches, the backshore may be missing, and the normal high-water uprush may impinge directly on cliffs or structures.

(b) Alternate terms for backshore are backbeach and berm. "Berm" is a common term because backshore areas are sometimes horizontal and resemble man-made berms. However, many beaches have a sloping backshore that does not resemble a berm, and some have more than one berm, representing the effects of several storms. Thus, berm is not synonymous with backshore, but may be a suitable description for selected areas. The term is sometimes used in beachfill and beach erosion control design.

(3) Coastline (or shoreline). The boundary between the foreshore and backshore, the high water line (hwl), is often defined to be the coastline. This is a practical definition because this land-water interface can be easily recognized in the field and can be approximated on aerial photographs by a change in color or shade of the beach sand (Crowell, Leatherman, and Buckley 1991). In addition, the coastline marked on the topographic sheets ("T-sheets") typically represents this same hwl, allowing a direct comparison between historic maps and aerial photographs. Some researchers have equated the coastline with the low-water line, but this boundary is not always marked by any evident feature or change in sand color. In various studies, one can find shoreline defined by almost any level datum. These inconsistencies make it difficult to compare shoreline maps prepared by different surveyors or agencies. Definition of "shoreline" often is controversial because it affects the legal definition of setback lines and other constraints placed on development in the coastal zone. A more detailed discussion of hwl identification is presented in Anders and Byrnes (1991); Crowell, Leatherman, and Buckley (1991); and Gorman, Morang, and Larson (1998).

*d. Beach material.*

(1) Sand beaches. On most of the coasts of the United States, the predominant beach material is sand (between 0.0625 and 2.0 mm, as defined by the Wentworth classification). Most sand beaches are composed mostly of quartz, with lesser percentages of feldspars, other minerals, and lithic (rock) fragments. Table IV-2-4 lists beach sediment types and common locations.

**Table IV-2-4**
**Types of Beach Sediment**

| Type | Typical Locations |
|---|---|
| Quartz sand | East Coast of U.S. between Rhode Island and North Florida, Gulf Coast between West Florida and Mexico, portions of West Coast of U.S. and Great Lakes |
| Calcite shell debris | South Florida, Hawaii |
| Volcanic sand | Hawaii, Aleutians, Iceland |
| Coral sand | South Florida, Bahamas, Virgin Islands, Pacific Trust Territory |
| Rock fragments | Maine, Washington, Oregon, California, Great Lakes |
| Clay balls | Great Lakes, Louisiana |

(2) Coarse beaches. Coarse beaches contain large amounts of granule-, pebble-, cobble-, and boulder-sized material (larger than 2.0 in the Wentworth classification). These beaches, found in the northeast, in the Great Lakes, and in mountainous reaches of the Pacific coast, occur under conditions where:

(a) Local streams flow with enough velocity to carry large particles to the shore.

(b) Coarse material underlies the beach (often found in areas influenced by glaciation or on metamorphic coasts).

(c) Coarse material is exposed in cliffs behind the beach (Figure IV-2-9).

The constituent material may be primarily angular rock fragments, especially if the source area, such as a cliff, is nearby (Figure IV-2-32). If the source area is far away, the most common rock types are likely to be quartzite or igneous rock fragments because these hard materials have a relatively long life in the turbulent beach environment. Softer rocks, such as limestone or shale, are reduced more readily to sand-sized particles

**Figure IV-2-32. Shale beach and bluffs, southeast shore of Lake Erie, near Evans, New York. Bedding planes in the shale are lubricated by groundwater, and freeze-thaw cycles split the rock along the planes. The rubble on the beach breaks down quickly, leaving a grey sand with plate-shaped grains**

by abrasion and breakage during their movement to the coast and by subsequent beach processes. Coarse beaches usually have a steeper foreshore than sand beaches.

(3) Biogenic beaches. In tropical areas, organically produced (biogenic) calcium carbonate in the form of skeletal parts of marine plants and animals can be an important or dominant constituent. The more common particles are derived from mollusks, barnacles, calcareous algae, Bryozoa, echinoids, coral, Foraminifera, and ostacods. The percentage of biogenic material in a beach varies with the rate of organic production and the amount of terrigenous material being contributed to the shore.

## IV-2-11. Salt Marshes

Coastal salt marshes are low-lying meadows of herbaceous plants subject to periodic inundations. During the constructional phase of a coastline, a marsh develops when sediment deposition exceeds sediment removal by waves. Three critical conditions are required for marsh formation: abundant sediment supply, low wave energy, and a low surface gradient. Once sediment accumulation reaches a critical height, the mud flats are colonized by halophytic plants that aid in trapping sediment when flooding occurs and add organic material to the substrate.

*a. Distribution of salt marshes.* Marshes occur in low-energy coastal locations, and the bay side of most barriers is fringed by tidal marsh or tidal flats. Likewise, mainland shorelines adjacent to bays are also typically fringed by marsh. Along the mainland of the United States, three regional marsh types have been recognized: (1) New England marsh; (2) Atlantic and Gulf Coastal Plain marsh; and (3) Pacific coast marsh

(Frey and Basan 1985). As much as 80 to 90 percent of the Atlantic and Gulf coasts have been bordering marsh around lagoons, estuaries, and deltaic environments (Inman and Nordstrom 1971). However, along the West Coast, less than 20 percent of the coastline has marsh because Pacific marshes are usually restricted to protected locations in river mouth lagoons or tectonically controlled bays. Morphologically and sedimentologically, Pacific marshes are similar to Atlantic and Gulf coast types, although they do differ in flora species and in the unequal semidiurnal range of tides.

New England marshes are adapted to high tide ranges, high wave energy, and cold winters. Atlantic Coastal Plain marshes are abundant and almost continuous from New York to northern Florida. The warm weather, low tide range, and low wave energy coastline of both the Atlantic and Gulf coastal areas of southern Florida gives rise to mangrove marshes. The deltaic plain of Louisiana supports a unique type of marsh called "flotant," which is marsh flora sitting atop organic rich ooze. The ooze is a soft substrate that may be several meters thick over harder substrates. The rest of the Gulf Coast has fringing marsh behind its barriers.

  b.  *Classification of salt marshes.*

  (1) Regional conditions such as temperature, sediment distribution, pH, Eh, and salinity contribute to the zonation of a marsh area. Plant successions, sediment accumulation, and marsh expansion vary but most marshes can be divided into two fundamental zones: low and high. Low marshes are younger, lower topographically, and usually subjected to the adjacent estuarine and marine processes. High marshes are older, occupy a higher topographic position, are more influenced by upland conditions, and are subjected to substantially fewer tidal submersions per year. The boundaries for these zones and their relationship to a given datum may differ from one coast to another. Differences in marsh boundaries might be related to tidal regularity and substrate composition. On the Atlantic coast, the tides are generally regular and near equal in semidiurnal range, whereas those on the Pacific coast are markedly unequal in semidiurnal range. Gulf Coast marshes are subjected to irregular and small-amplitude tides. Consequently, the demarcation of high and low marshes is not well-defined.

  (2) Plant structures and animals are significant contributors to sediment accumulation in salt marshes (Howard and Frey 1977). Grasses dampen wind-generated waves. Stems and levees impede current flow, which helps trap suspended sediment (Deery and Howard 1977). The most obvious mechanism of sediment entrapment is the plant root system. Plant roots may extend more than a meter in depth along Georgia stream-side marshes and up to 50 cm in some adjacent habitats (Edwards and Frey 1977).

  c.  *Sediment characteristics.*

  (1) Introduction.

  (a) Salt marshes generally contain finer, better-sorted sediment than other intertidal environments. However, marsh substrates reflect the local and regional sediment sources. Along the Atlantic coast and shelf of the United States, Hathaway (1972) recognized two distinct clay mineral facies. The northern clay-mineral facies, extending from Maine to Chesapeake Bay, is primarily composed of illite, chlorite, and traces of feldspar and hornblende. The southern clay-mineral facies, which extends from Chesapeake to the South, is composed of chiefly kaolinite and montmorillonite.

  (b) In New England and along many northern coasts overseas, peat is an important soil component of marsh substrate. Peat forms from the degradation of roots, stems, or leaves of marsh plants, particularly *Spartina* (Kerwin and Pedigo 1971). In contrast, peat is not a significant component of the southern coastal marshes except in Louisiana and Florida (Kolb and van Lopik 1966). The southern marsh substrate generally consists of silt- and clay-size sediment with a large percentage of carbon material. The major sources of organic carbon in most coastal marshes are in-situ plants and animal remains.

(c) Southern marshes can have up to 60 percent silt and up to 55 percent clay. Rapid biological decay and constant flushing prevent the accumulation of thick organic deposits. The exception to this is the "flotant" marshes, which may have over 50 percent organics. In the Unified Soil Classification, the substrate would be considered a fine-grained soil, and the field engineer could anticipate silt, clay, or organic-dominant material with low to high plasticity. Much of the organics present in marshes are incompletely oxidized, which if released into the lagoon/bay by excavation, can profoundly affect the water chemistry.

(2) Marsh plants.

(a) Marsh plants are typically tall, salt-tolerant grasses. About 20 genera of salt marsh plants are found worldwide, with the most important in North America being *Spartina*, *Juncus*, and *Salicornia* (Chapman 1974). Salt marshes are the temperate (and arctic) counterparts of tropical mangrove forests. They generally develop in shallow, low-energy environments where fine-grained sediments are deposited over sandy or till substrates (Figure IV-2-33). As the fine sediments build upward, the marsh plants can take root and become established. The established vegetation increases sediment trapping and leads to more rapid upward and outward building of marsh hummocks, which form the foundation of the marsh. The vegetation also creates lower energy conditions by absorbing wave energy and reducing current velocities, thus allowing accelerated sediment deposition.

(b) Like mangrove forests, many species of invertebrates, fish, birds, and mammals inhabit salt marshes and the adjacent tidal creeks during all or part of their life cycles. Thus, marshes are important to commercial and sport fishermen and hunters. In addition, several marsh species are considered endangered.

(c) Also like mangrove forests, man's main detrimental impact on these marshes has been dredge-and-fill operations for land reclamation and mosquito control. Air and water pollution are also serious problems. Although extensive areas of salt marsh remain on the east and Gulf coasts of North America, significant areas have been lost to development. The situation is much worse on the west coast, where most of the coastal marsh lands have been filled and perhaps permanently destroyed. Efforts to restore degraded coastal marshes have not generally been successful.

(3) Sediment transport and processes.

(a) Typically, most marshes have very slow rates of sediment accumulation, amounting to only a few millimeters per year (Pethick 1984). Natural and man-induced changes can have deleterious effects on marsh growth. For example, building levees or altering the drainage pattern can result in erosion and permanent marsh loss. Not only is suspended sediment important to vertical growth of the marsh, but biologic components, particularly organic detritus suspended in the water column, are critical to marsh health. The exchange of sediment and nutrients is dependent on the exchange between the local bodies of water.

(b) A marsh sediment budget usually includes consideration of the following factors (Davis 1985):

- Riverine sources.

- Offshore or longshore transport.

Figure IV-2-33. Estuary of the Sprague River, Maine (a small river facing the Gulf of Maine west of Popham Beach State Park). This is a typical New England salt marsh, with sinuous channels that alternately flood and drain depending on the stage of the tide. In the past, straight channels were often dug in such marshes to allow more efficient draining to reduce mosquito infestation

- Barrier washover.

- Headlands.

- Eolian transport.

- In situ organic material (i.e., peat, plant detritus, and feces).

- Other terrestrial sources.

(4) Engineering problems. In light of growing concerns to preserve natural coastal marshes and the need to implement the national policy of "no net wetland losses," many agencies are researching ways to manage and implement wetland technology. Studies have identified numerous man-made and natural causes of wetland loss in the coastal zone:

(a) Sediment deficits. Man-made modifications of natural fluvial systems interfere with natural delta-building processes.

(b) Shoreline erosion. Along many marsh shorelines, the rates of retreat have increased because of hurricanes and other storms, engineering activities along the coast, and boating.

(c) Subsidence. Sinking of the land due to natural compaction of estuarine, lagoonal, and deltaic sediments results in large-scale disappearance of wetlands. This effect is exacerbated in some areas (e.g., Galveston Bay) by subsidence caused by groundwater and oil withdrawal.

(d) Sea level rise. Eustatic sea level rise is partially responsible for increased rates of erosion and wetland loss.

(e) Saltwater intrusion. Increased salinities in wetlands harm vegetation, which makes the wetlands more vulnerable to erosion.

(f) Canals. Canals increase saltwater intrusion and disrupt the natural water flow and sediment transport processes.

(5) Marsh restoration. Many agencies, including the U.S. Army Corps of Engineers, are conducting research in the building and restoration of marshes, are developing marsh management techniques, and are developing regulatory guidelines to minimize land loss. Under the Wetlands Research Program sponsored by the USACE, new technology in a multidisciplined approach is being developed. A useful publication is the "Wetlands Research Notebook" (U.S. Army Engineer Waterways Experiment Station 1992), which is a collection of technical notes covering eight field problem areas focusing on wetlands activities in support of USACE civil works projects.

## IV-2-12. Biological Coasts

*a. Introduction.*

(1) On many coasts, such as open wetlands, coral reef, and mangrove forest, biological organisms and processes are of primary importance in shaping the morphology. In contrast, on many other coasts, such as typical sandy beaches, biological activities do not appear to be of major significance when compared to the physical processes at work. Nevertheless, it is important to realize that biological processes are occurring on all shores; all man-made shoreline modifications must address the impact of the modification on the biological community.

(2) The types of organisms that can exist on a coast are ultimately controlled by interrelated physical factors. These include wave climate, temperature, salinity, frequency of storms, light penetration, substrate, tidal range, and the amounts of sediments and nutrients available to the system. Of these, the most important may be wave climate. The amount of wave energy dissipated at a shoreline per unit time ultimately has a dominant influence on whether the substrate is rock, sand, or silt; on the water clarity; on the delivery of nutrients; and, most importantly, on an organism's physical design and lifestyle. The physical forces exerted by a large breaking wave are several orders of magnitude greater than the typical lateral forces affecting organisms in most other environments. For example, mangroves and salt marshes require low wave-energy climates to provide suitable substrate and to keep from being physically destroyed. On the other hand, reef-building corals require reasonably high wave-energy environments to maintain the water clarity, to deliver nutrients, to disperse larvae, to remove sediment, and to limit competition and predation.

(3) Another first-order physical condition controlling biological organisms is temperature. For example, this is the primary factor that restricts mangroves and coral reefs to the tropics. Also, the formation of ice in coastal waters has a major impact on Arctic communities.

(4) Unlike many physical processes on coastlines, biological processes are generally progradational in nature, extending shorelines seaward. Reef-building organisms produce hard substrate and sediments, in addition to sheltering areas behind the reefs. Some mollusks, calcareous algae (*Hallemeda sp.*, etc.), barnacles, echinoids, bryozoa, and worms produce significant amounts of sediment. Under low-energy

conditions in the deep sea and sheltered waters, diatoms and radiolaria produce sediments. Mangroves, salt marsh, and dune vegetation trap and stabilize sediments. The erosional effect of organisms that burrow into sediments or that bore into rocks is usually of lesser importance.

  *b. High wave-energy coasts.* Higher plants have not evolved mechanisms to enable them to physically withstand high wave-energy environments. Thus, simple plants, mainly algaes, form the bases of the food chains for these marine coastal communities.

  (1) Coral reefs. Coral reefs are massive calcareous rock structures that are slowly secreted by simple colonial animals that live as a thin layer on the rock surface. The living organisms continually build new structures on top of old, extending the reefs seaward toward deeper water and upward toward the surface. Reef-building corals have algae living within their tissues in a symbiotic relationship. The algae supplies food to the coral and the coral supplies shelter and metabolic wastes as nutrients to the algae. Shallow coral reefs worldwide occupy some 284,300 km$^2$, about 1.2 percent of the world's continental shelf area (Spalding, Ravilious, and Green 2001). While some corals are found in temperate and Arctic waters, reef-building corals are limited by water temperature to the tropics, mainly between the latitudes of 30 deg north and south. Bermuda, in the North Atlantic, warmed by the Gulf Stream, is the highest latitude location where active coral reefs are presently found. In the United States, coral reefs are found throughout the Florida Keys and the east and west coasts of Florida, in the Hawaiian Islands, the Pacific Trust Territories, Puerto Rico, and the Virgin Islands.

  (a) Reef-building corals require clear water. The corals need to be free of sediments in order to trap food particles, and their algae require sufficient light for photosynthesis. While corals can remove a certain amount of sediment from their upper surfaces, heavy siltation will bury and kill them. Light penetration limits the depth of most reef-building corals to the upper 30-50 m, though some corals grow much deeper. The upper limit of reef growth is controlled by the level of low tide. Corals cannot stand more than brief exposures out of the water (for example, during the occasional passage of a deep wave trough).

  (b) While coral reefs produce rock structure, they also produce calcareous sediments. Waves and currents pulverize coral skeletons into sand-size particles. However, on many reefs, calcareous algae (Hallemeda sp.) produce a majority of the sediments. The crushed calcareous shells of other animals, such as mollusks, sea urchins, and sand dollars, also provide sediment.

  (c) Coral reefs rival tropical rain forests as being among the most complex communities on earth, and rock-producing reef communities are among the most ancient life forms found in the fossil record. Because of their complexity, the dynamics of coral reefs are not yet well understood. At least 100,000 species have been named and described, but the total number inhabiting the world's reefs may exceed one million species. Scientific knowledge of the ecology of reefs has almost entirely accumulated over the last 50 years, since the development of underwater scuba breathing apparatus in the 1940's.

  (d) One of the critical ecological issues of our times is the rapid degradation of coral reefs around the world by various natural and human activities. Corals are highly sensitive to increases in temperature, exhibiting a stress response known as coral bleaching. In 1998, a global mass bleaching caused mass mortalities in many areas. Worldwide, however, humans are driving more profound changes to reefs than are natural phenomena. The most widespread impacts are water pollution, dredge and fill operations, over-harvesting of fish and shellfish, and the harvesting of some corals for jewelry. For example, in Indonesia, the world's largest coral nation, 82 percent of the corals are at risk from the illegal practice of dynamite fishing (which is also devastating to fish populations). Even far from the coast, deforestation, urban sprawl, and sloppy agriculture produce vast quantities of sediment and pollution that enter the sea and degrade reefs in the vicinity of river mouths.

(e) Controlled dredging around reefs is possible and is done routinely, causing minimal impact to reef communities. Mechanical damage (from cutter heads, chains, anchors, and pipelines) is often of equal or greater concern than suspended sediment production. Improvements in navigation and positioning have made dredging near reefs more viable. Nevertheless, careful monitoring is mandated in most cases.

(f) Reefs are of major economic importance to the communities along which they are located. For millenia, coastal peoples relied on coral reefs as a source of food. Spurgeon (1992) classifies their economic benefits as:

* Direct extractive uses - fisheries, building material.

* Direct non-extractive uses - tourism.

* Indirect uses - biological support for a variety of other ecosystems.

Reef tourism is now a major global industry. Diving tourism is ubiquitous and now occurs in 91 countries and states (Spalding, Ravilious, and Green 2001). One major benefit of mass tourism is that it had brought to public consiousness the issues of the fragility of reef ecosystems and the need to preserve the world's remaining reefs. Marine protected areas are becoming a critical tool in the preservation of remaining reefs, and some 660 protected areas worldwide incorporate reefs. Tourist income is helping some remote communities police and protect their reefs. The downside of mass tourism is the haphazard growth of infrastructure along the shore. Many vacation communities are built on remote coasts where resources, such as fresh water, are scarce and where trash and sewage effluent are not properly managed.

(g) An additional benefit of reefs is the shelter from waves that they provide to adjacent shores. As an example, the south and southwestern coast of Sri Lanka is battered by waves that travel unhindered across the Indian Ocean. In the past, the coral reefs that surrounded the coast served as buffers against the intense wave energy. But illegal reef breaking and coral mining, combined with negative impacts of tourism and development (sewage, agricultural pollution, physical damage) have greatly reduced the effectiveness of the reef barriers. As a result, much of the Sri Lankan south coast is now experiencing severe erosion (Young and Hale 1998).

(h) Stoddard (1969) has identified four major forms of large-scale coral reef types: fringing reefs, barrier reefs, table reefs, and atolls.

(i) *Fringing reefs* generally consist of three parts: a fore reef, a reef crest, and a back reef. The *fore reef* usually rises steeply from deep water. It may have spur and groove formations of coral ridges interspersed with sand and rubble channels. The *reef crest* usually forms a continuous wall rising to the low tide level. This usually occurs within a few hundred meters from shore. The seaward side of this area, called the *buttress zone*, receives the brunt of the wave action. Between the reef crest (or flat) and the shoreline, the reef usually deepens somewhat in the back reef area. This area typically contains much dead coral as well as rock, rubble, sand, and/or silt. It also contains live coral heads, algae, eel grass, etc. Fringing reefs form as the beginning stages in the evolution of atolls and possibly barrier reefs.

(j) *Barrier reefs* grow on the continental shelf where suitable solid substrate exists to serve as a foundation. Their form is typically a long coral embankment separated from the mainland by a lagoon that may be several kilometers wide. The lagoon is usually flat-floored and may be as much as 16 km wide and 35 to 75 m in depth. Although similar to fringing reefs, barrier reefs are much more massive, the reef crests are much further from shore, and the back reef areas are deeper. Protected shorelines behind barrier reefs are characterized by mangrove swamps and are usually progradational. The seafloor on the seaward side slopes steeply away into deeper water and is covered by coral rubble.

(k) *Table reefs* grow from shallow banks on the seafloor that have been capped with reef-forming organisms. They cover extensive areas but do not form barriers or enclose lagoons.

(l) *Atolls* are ring-shaped reefs that grow around the edges of extinct volcanic islands, enclosing lagoons of open water. The shallow lagoons may contain patch reefs. Atolls are primarily found in isolated groups in the western Pacific Ocean. Small low islands composed of coral sand may form on these reefs. These islands may hold enough of a fresh water lens to support human life, but the islands are quite vulnerable to inundation and to tropical storms. The first theory concerning the development of atolls, the subsidence theory proposed by Charles Darwin in 1842, has been shown to be basically correct (Strahler 1971). Figure IV-2-34 illustrates the evolution of an atoll.

(m) The development of atolls begins with an active volcano rising from the ocean floor and forming a volcanic island. As the volcano ceases activity, a fringing reef forms along the shore. Over geologic time, erosion of the volcanic island and subsidence due to general aging of the ocean basin cause the island to drop below sea level. The actively growing fringing reef keeps pace with the subsidence, building itself upward until a barrier reef and lagoon are formed. As the center of the island becomes submerged, the reef continues its upward growth, forming a lagoon. During the development, the lagoon floor behind the reef accumulates coral rubble and other carbonate sediments, which eventually completely cover the subsiding volcanic island.

(2) Worm reefs. A type of biogenic reef that is not related to coral reefs is that produced by colonies of tubeworms. Serpulid worms and Sabellariid worms are two types known to form significant reef structures by constructing external tubes in which they live. The Serpulids build their tubes from calcareous secretions and the Sabellariids by cementing particles of sand and shell fragments around their bodies. Colonies of these worms are capable of constructing massive structures by cementing their tubular structures together. As new tubes are continually produced over old ones, a reef is formed. These reefs typically originate from a solid rocky bottom, which acts as an anchoring substrate. Worm reefs are most commonly found in sub-tropical and tropical climates (e.g., east coast of Florida). Reefs of this nature can play an important role in coastal stabilization and the prevention of coastal erosion.

(3) Oyster reefs. Oysters flourish under brackish water conditions in lagoons, bays, and estuaries. The oysters cement their shells to a hard stable substrata including other oyster shells. As new individuals set onto older ones, a reef is formed. These reefs can form in temperate as well as tropic waters.

(a) Oysters found around the United States are part of the family *Ostreidae*. The Eastern, or American oyster (*Crassostrea virginica*) is distributed along the entire east coast of North America from the Gulf of St. Lawrence through the Gulf of Mexico to the Yucatan and the West Indies. The other major North American species is *Ostrea lurida*, which ranges along the Pacific coast from Alaska to Baja California (Bahr and Lanier 1981).

(b) Intertidal oyster reefs range in size from isolated scattered clumps a meter high to massive solid mounds of living oysters anchored to a dead shell substrate a kilometer across and 100 m thick (Pettijohn 1975). Reefs are limited to the middle portion of the intertidal zone, with maximum elevation based on a minimum inundation time. The uppermost portion of a reef is level, with individual oysters pointing upwards. At the turn of the century, vast oyster flats were found along the Atlantic coast in estuaries and bays. In South Carolina, the flats covered acres and sometimes square miles (cited in Bahr and Lanier (1981)).

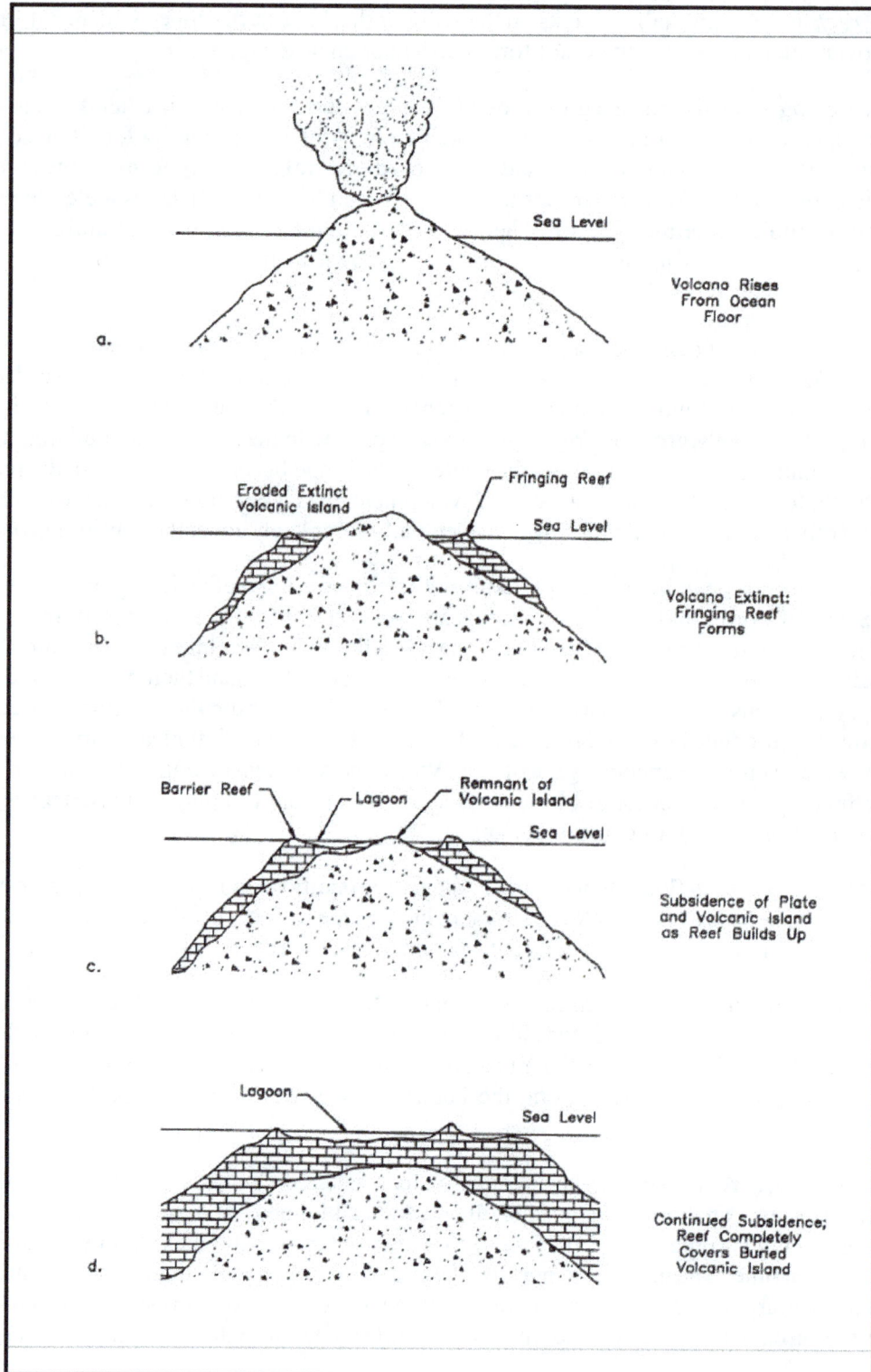

Figure IV-2-34. Evolution of a coral island: (a) Active volcano rising from the seafloor, (b) Extinct volcanic island with fringing reef, (c) Subsiding island; reef builds upward and seaward, forming barrier reef, (d) Continued subsidence causing remnant volcanic island to be completely submerged. Growth continues upward and seaward until remnant volcano is covered (adapted from Press and Siever (1986))

(c) Oyster reefs serve an important biological role in the coastal environment. The reefs are a crucial habitat for numerous species of microfauna and macrofauna. The rough surface of a reef flat provides a huge surface area for habitation by epifauna, especially vital in the marsh-estuarine ecosystem that is often devoid of other hard substrate. The high biological productivity of reef environments underscores one of the reasons why reefs must be protected and preserved.

(d) Oyster reefs play important physical and geological roles in coastal dynamics because they are wave-resistant structures that can biologically adapt to rising sea level. Reefs affect the hydrologic regime of salt marsh estuaries in three ways: by modifying current velocities, by passively changing sedimentation patterns, and by actively augmenting sedimentation by biodeposition. (Biological aggradation increases the size of suspended particles and increases their settling rates.) As reefs grow upward and laterally, modify energy fluxes by damping waves and currents, and increase sedimentation, they ultimately produce major physiographic changes to their basins. These changes can occur on short time scales, on the order of hundreds of years (Bahr and Lanier 1981). During geologic history, massive reefs have accumulated in many areas, some of which became reservoirs for oil and gas.

(e) Although oysters are adapted to a wide range of temperature, turbidity, and salinity conditions, they are highly susceptible to man-made stresses. These stresses on oyster communities can be classified into eight categories (Bahr and Lanier 1981):

- Physical sedimentation, especially from dredging or boat traffic.

- Salinity changes due to freshwater diversion or local hydrologic alterations.

- Eutrophication (oxygen depletion) due to algae growth in water that is over-enriched with organic matter.

- Toxins from industrial and urban runoff.

- Physical impairment of feeding structures by oil.

- Thermal loading, primarily from power plants.

- Overharvesting.

- Loss of wetlands.

There has been a recorded significant decline in the health and extent of living United States east coast oyster reefs since the 1880's, although the data are sometimes conflicting because ground-level surveys are difficult to conduct (Bahr and Lanier 1981). It is easy to account for the declines of reefs near population and industrial centers, but the declines are more difficult to explain in more pristine areas of the coast (e.g. the Georgia coast near Sapelo Island). Population changes may be due, in part, to natural cycles of temperature and salinity or fecundity.

(f) Because oyster reefs are susceptible to fouling and silting, it is important that geologists and engineers consider sediment pathways during the planning phases of coastal construction and dredging projects or stream diversion and other watershed changes. As discussed earlier, dredging near reefs is technically feasible as long as careful technique is observed and environmental conditions are monitored.

(g) In summary, oyster reefs serve critical biological and physical purposes in the estuarine and coastal marsh environment. They enhance biological productivity, provide stable islands of hard substrate in otherwise unstable soft muddy bottoms, modify hydrodynamic flows and energy fluxes. With respect to shore protection, reefs are a biological wave damper that can accommodate rising sea level as long as they are alive. It is essential that reefs be protected from wanton destruction by pollution and other stresses imposed by human development.

(4) Rocky coasts.

(a) Kelp beds. Kelp forests are formed by various species of algae that attach to hard substrate with a root-like system called a *holdfast*. Some (prominently *Macrosistus sp.*) can grow many tens of meters in length up to the water surface, where their tops float and continue to grow. The plants are quite rubbery and can withstand significant wave action. Kelp beds are found along rocky shorelines having cool clear water. In North America, they occur along much of the Pacific coast and, to a lesser extent, along the North Atlantic coast. Kelp beds are, to some extent, the functional temperate latitude counterpart of coral reefs (Carter 1988).

(b) Kelp biological communities. Kelp beds harbor extensive biological communities that include fish, sea otters, lobster, starfish, mollusks, abalones, and many other invertebrates. In addition, kelp beds absorb wave energy, helping to shelter beaches. Man's main impact has been the commercial harvesting of various portions of this community, including the kelp. In the past, hunting sea otters for their pelts allowed sea urchins to multiply, and the overpopulation of sea urchins grazed and destroyed many beds. Today, the reestablishment of some sea otter populations has led to conflicts with shell fishermen. Water pollution is also a problem in some areas.

(c) Rock reefs and shorelines. Submerged rock reefs provide substantial habitat for organisms. They provide a place of attachment for sessile organisms, and the crevices provide living spaces and havens of refuge for mobile organisms such as fish and lobsters. These structures are a boon to sports fishermen, and many artificial reefs have been built on sandy seafloors out of a wide array of materials. Rocky shorelines have communities of organisms living in the intertidal and subtidal zones. These may or may not be associated with offshore kelp bed or coral reef communities.

(5) Sandy coasts. Much of the biological activity on sandy coasts is confined to algae, various inverte-brates, and fish living within the water column. Of these, fish, shrimp, and crabs have the greatest economic importance. In addition, there are infaunal filter feeders, mainly mollusks and sand dollars, that live just beneath the sand surface.

(a) One important and often overlooked biological activity on some sandy beaches is their use as nesting areas by a variety of migratory animals. These include sea turtles, birds, marine mammals, and fish. In North America, many of these species are threatened or endangered, including all five species of sea turtles and some birds such as the piping plover, the snowy plover, and the least tern. For most of these species, their problems are directly related to conflicts with man's recreational use of beaches and the animals' inability to use alternate nesting sites. Fortunately, some states have implemented serious ecological programs to help save these threatened species. For example, Florida has rigorous laws preventing disruption of nesting turtles, and many Florida municipalities have found that maintaining healthy natural biological communities is an excellent way to lure tourists.

(b) Plants occupying sand dunes are characterized by high salt tolerance and long root systems that are capable of extending down to the freshwater table (Goldsmith 1985). Generally, these plants also generate rhizomes that grow parallel to the beach surface. Beach plants grow mainly in the back beach and dune areas beyond the zone of normal wave uprush. The plants trap sand by producing low energy conditions near the ground where the wind velocity is reduced. The plants continue to grow upward to keep pace with the accumulation of sand, although their growth is eventually limited by the inability of the roots to reach dependable water. The roots also spread and extend downward, producing a thick anchoring system that stabilizes the back beach and dune areas. This stabilization is valuable for the formation of dunes, which provide storm protection for the entire beach. The most common of these plants are typically marram grass, saltwort, American sea grass, and sea oats. With time, mature dunes may accumulate enough organic nutrients to support shrub and forest vegetation. The barrier islands of the U.S. Atlantic coast and the Great Lakes shores support various species of *Pinus*, sometimes almost to the water's edge.

*c. Low wave-energy coasts.* In locations where the wave climate is sufficiently low, emergent vegetation may grow out into the water. Protection from wave action is typically afforded by local structures, such as headlands, spits, reefs, and barrier islands. Thus, the vegetation is usually confined to the margins of bays, lagoons, and estuaries. However, in some cases, the protection may be more regional in extent. Some of the mangrove forests in the Everglades (south Florida) and some of the salt marshes in northwest Florida and Louisiana grow straight into the open sea. The same is true for freshwater marshes in bays and river mouths in the Great Lakes.

(1) General.

(a) Only a few higher plants possess a physiology that allows them to grow with their roots in soils that are continuously saturated with salt water. These are the mangroves of the tropics and the salt marsh grasses of the higher latitudes. The inability of other plants to compete or survive in this environment allows small groups of species or single species to cover vast tracts of some coastal areas. These communities typically show zonations with different species dominant at slightly different elevations, which correspond to different amounts of tidal flooding. The seaward limit of these plants is controlled by the need for young plants to have their leaves and branches above water. To this end, some mangroves have seedlings that germinate and begin growing before they drop from the parent tree. Upland from these communities, a somewhat larger number of other plants, such as coconuts and dune grasses, are adapted to live in areas near, but not in, seawater.

(b) Understanding and appreciation of the importance of these types of coastal areas are growing. Former attitudes that these areas were mosquito-infested wastelands imminently suitable for dredge- and-fill development are being replaced by an appreciation of their great economic importance as nursery grounds for many species of fish and shellfish, of their ability to remove pollutants, of their ability to protect upland commercial development from storms, of their fragility, and of their beauty.

(2) Mangroves.

(a) *Mangroves* include several species of low trees and shrubs that thrive in the warm, shallow, saltwater environments of the lower latitudes. Worldwide, there are over 20 species of mangroves in at least 7 major families (Waisel 1972). Of these, the red, white, and black mangroves are dominant in south Florida and the Caribbean. They favor conditions of tidal submergence, low coastal relief, saline or brackish water, abundant fine sediment supply, and low wave energy. Mangroves have the ability to form unique intertidal forests that are characterized by dense entangled networks of arched roots that facilitate trapping of fine sediments, thereby promoting accretion and the development of marshlands. The *prop roots* and *pneumatophores* also allow the plants to withstand occasional wave action and allow oxygen to reach the roots in anaerobic soils. The prime example in the United States is the southwest shore of Florida, in the Everglades National Park.

(b) Mangrove coasts are crucial biological habitats to a wide variety of invertebrates, fish, birds, and mammals. In the past, the primary cause of their destruction has been dredge-and-fill operations for the reclamation of land and for mosquito control.

*d. Other sources of biogenic sediment in the coastal zone.* In areas of high biological activity, organically derived sediments may account for a significant proportion of the sediment composition, especially where terrigenous sediment supplies are low. These biogenic materials, consisting of remains of plants and animals and mineral matter produced by plants and animals, accumulate at beaches, estuaries, and marshlands.

(1) The most familiar types of biogenic sediments are hard calcareous skeletal parts and shell fragments left behind by clams, oysters, mussels, corals, and other organisms that produce calcareous tests. In tropical climates, the sediment commonly consists of coral fragments and calcareous algal remains. Siliceous tests are produced by most diatoms and radiolaria. Sediments predominately containing carbonate or calcareous material are generally referred to as *calcarenites*, while sediments composed predominantly of siliceous matter are referred to as *diatomites* or *radiolarites*, depending upon which organism is most responsible for the sediment (Shepard 1973). In the Great Lakes and some inland U.S. waterways, the zebra mussel has proliferated since the mid-1980's, and now some shorefaces are covered with mussel shell fragments to a depth of over 10 cm. The mussels are a serious economic burden because they choke the inlets of municipal water systems and coolant pipes.

(2) In some areas, wood and other vegetation may be introduced into the sediments in large quantities. This is especially common near large river mouths and estuaries. This organic material may become concentrated in low energy environments such as lakes and salt marshes, eventually producing an earthy, woody composition known as *peat* (Shepard 1973). Peat exposed on the shoreface has been used as an indicator of marine transgression and barrier island retreat (Figure IV-2-35) (Dillon 1970). In Ireland and Scotland, peat is dried and used as a fuel.

## IV-2-13. Continental Shelf Geology and Topography

*a. Introduction.* The geology of the world's continental shelves is of direct significance to coastal engineers and managers in two broad areas. First, the topography of the shelf affects coastal currents and wave climatology. Wave refraction and circulation models must incorporate shelf bathymetry. Bathymetry was incorporated in the wave hindcast models developed by the USACE Wave Information Study (See references in Part II-8). Second, offshore topography and sediment characteristics are of economic importance when offshore sand is mined for beach renourishment or dredged material is disposed offshore.

*b. Continental shelf sediment studies.*

(1) The Inner Continental Shelf Sediment and Structure (ICONS) study was initiated by the Corps of Engineers in the early 1960's to map the morphology of the shallow shelf and find sand bodies suitable for beach nourishment. This program led to a greater understanding of shelf characteristics pertaining to the supply of sand for beaches, changes in coastal and shelf morphology, longshore sediment transport, inlet migration and stabilization, and led to a better understanding of the Quaternary shelf history. ICONS reports are listed in Table IV-2-5.

**Figure IV-2-35.** Peat horizon exposed on the shoreface, Ditch Plains, Long Island, New York (March 1998). The peat is in-situ, indicating that lagoonal sediments accumulated here before the barrier beach retreated over the marsh. The peat layer was about 1 m above the ocean water level at the time the photograph was taken. The dune resting on the peat is about 2 m thick

(2) Since the 1970's, the Minerals Management Service (MMS), a bureau of the U.S. Department of the Interior, has been tasked with managing the mineral resources of the Outer Continental Shelf. In conducting its mission, the MMS has sponsored many surveys and studies of mineral resources on the continental shelf. These studies can be accessed via the MMS' Environmental Studies Program Information System (ESPIS) at:

http://mmspub.mms.gov/espis/

(3) For beach renourishment projects, U.S. Army Corps of Engineers Districts typically obtain information on sand resources near the proposed project area from various sources:

(a) In-house studies, typically from vibracoring or rotary borings.

(b) Contracts with marine geophysics/geotechnical surveyors.

(c) The U.S. Geological Survey.

**Table IV-2-5**
**U.S. Army Corps of Engineers Inner Continental Shelf Sediment and Structure (ICONS) Reports**

| Location | References[1] |
| --- | --- |
| **Atlantic Coast** | |
| Massachusetts Bay | Meisburger 1976 |
| New York - Long Island Sound | Williams 1981 |
| New York - Long Island shelf | Williams 1976 |
| New York Bight | Williams and Duane 1974 |
| New Jersey - central | Meisburger and Williams 1982 |
| New Jersey - Cape May | Meisburger and Williams 1980 |
| Delaware, Maryland, Virginia | Field 1979 |
| Chesapeake Bay entrance | Meisburger 1972 |
| North Carolina - Cape Fear | Meisburger 1977; Meisburger 1979 |
| Southeastern U.S. shelf | Pilkey and Field 1972 |
| Florida - Cape Canaveral to Georgia | Meisburger and Field 1975 |
| Florida - Cape Canaveral | Field and Duane 1974 |
| Florida - Palm Beach to Cape Kennedy | Meisburger and Duane 1971 |
| Florida - Miami to Palm Beach | Duane and Meisburger 1969 |
| **Gulf of Mexico** | |
| Texas - Galveston County | Williams, Prins, and Meisburger 1979 |
| **Lake Erie** | |
| Pennsylvania | Williams and Meisburger 1982 |
| Ohio | Williams et al. 1980; Carter et al. 1982 |
| **Lake Michigan** | |
| Southeast shore | Meisburger, Williams, and Prins 1979 |
| **Sampling tools and methods** | |
| Pneumatic coring device | Fuller and Meisburger 1982 |
| Vibratory samplers | Meisburger and Williams 1981 |
| Data collection methods | Prins 1980 |

[1] Complete citations are listed in Appendix A

(d)  U.S. Army Engineer Research and Development Center, Waterways Experiment Station.

(e)  State and local agencies.

New geographic data collected by the Federal Government is documented with Metadata that can be accessed from various computer servers. Unfortunately, there is no consistent method of cataloging historical data or reports. Users who need information on sand resources near the coast must contact the Corps District responsible for that particular area.

*c. Continental shelf morphology.* Surficial sediment on the continental shelves is largely dependent upon the type of coast (i.e. collision, or leading, versus trailing) and the presence of rivers that supply material to the coast.

(1) Leading edge shelves, such as the Pacific coasts of North and South America, are typically narrow and steep. Submarine canyons, which sometimes cut across the shelves almost to the shore (Shepard 1973), serve as funnels that carry sediment down to the abyssal plain. Normally, very little sand is available offshore.

(2) Trailing edge shelves are, in contrast, usually wide and flat, and the heads of canyons usually are located a considerable distance from shore. Nevertheless, a large amount of sediment is believed to move down these canyons (Emery 1968). Off the United States Atlantic coast, the broad continental shelf contains a vast amount of sand. Unfortunately, much of this sand is not available for beach renourishment because it is either too far from shore or its composition is incompatible with the beaches where it is to be placed (i.e., contains too much rock, shell fragment, mud, or organic material or the grain size is different than the size of the native material where it is to be placed).

*d. Examples of specific features - Atlantic seaboard.*

(1) The continental shelf of the Middle Atlantic Bight of North America, which is covered by a broad sand sheet, is south of the region directly influenced by Pleistocene glacial scouring and outwash. This sand sheet is divided into broad, flat, plateau-like compartments dissected by shelf valleys that were excavated during the Quaternary lowstands of the sea. Geomorphic features on the shelf include low-stand deltas (cuspate deltas), shoal and cape retreat massifs (bodies of sand that formed during a transgressive period), terraces and scarps, cuestas (asymmetric ridges formed by the outcrop of resistant beds), and sand ridges (Figure IV-2-36) (Swift 1976; Duane et al. 1972).

(2) The larger geomorphic features of the Middle Atlantic Bight are constructional features molded into the Holocene sand sheet and altered in response to storm flows. Off the coasts of Delaware, Maryland, and Virginia, shoreface-connected shoals appear to have formed in response to the interaction of south-trending, shore-parallel, wind-generated currents with wave- and storm-generated bottom currents during winter storms. Storm waves aggrade crests, while fair-weather conditions degrade them. A second shoal area further offshore at the 15-m depth is indicative of a stabilized sea level at that elevation. These shoals may be suitable sources of sand for beach renourishment. However, the often harsh wave conditions off the mid-Atlantic seaboard may limit the economic viability of mining these shoals. The origin and distribution of Atlantic inner shelf sand ridges are discussed in McBride and Moslow (1991).

(3) Linear shoals of the Middle Atlantic Bight tend to trend northeast (mean azimuth of 32 deg) and extend from the shoreline at an angle between 5 and 25 deg. Individual ridges range from 30 to 300 m in length, are about 10 m high, and have side slopes of a few degrees. The shoal regions extend for tens of kilometers. The crests are composed of fine-medium sand, while the ridge flanks and troughs are composed of very fine-fine sand. The mineralogy of shoals reflects that of the adjacent beaches.

*e. Riverine influence.*

(1) Rivers provide vast amounts of sediment to the coast. The 28 largest rivers of the world, in terms of drainage area (combined size of upland drainage area and subaerial extent of deltas), discharge across trailing-edge and marginal sea coasts (Inman and Nordstrom 1971). Because the larger rivers drain onto trailing edge coasts, these shores tend to have larger amounts of available sediment, which is deposited across a wide continental shelf. The sediment tends to remain on the shelf and is only lost to the abyssal plains when deltas prograde out across the continental rise (e.g., the Mississippi and Nile Deltas) or when submarine canyons are incised across the shelf (e.g., Hudson River sediment funnels down the Hudson Canyon).

(2) The Columbia River, which is the 29th largest river in the world, is the largest one to drain across a collision coast. Until dams were built during the mid-20th century, the Columbia carried a major sediment load, which was deposited on the ebb shoal off its mouth. This shoal provided the sand that formed the Long Beach peninsula and fed the beaches as far north as the Olympic Peninsula. The Columbia appears to be an exception - on most collision coasts, canyons frequently cut across the shelf almost to the shore (Shepard 1973), therefore resulting in the direct loss of sediment from the coastal zone.

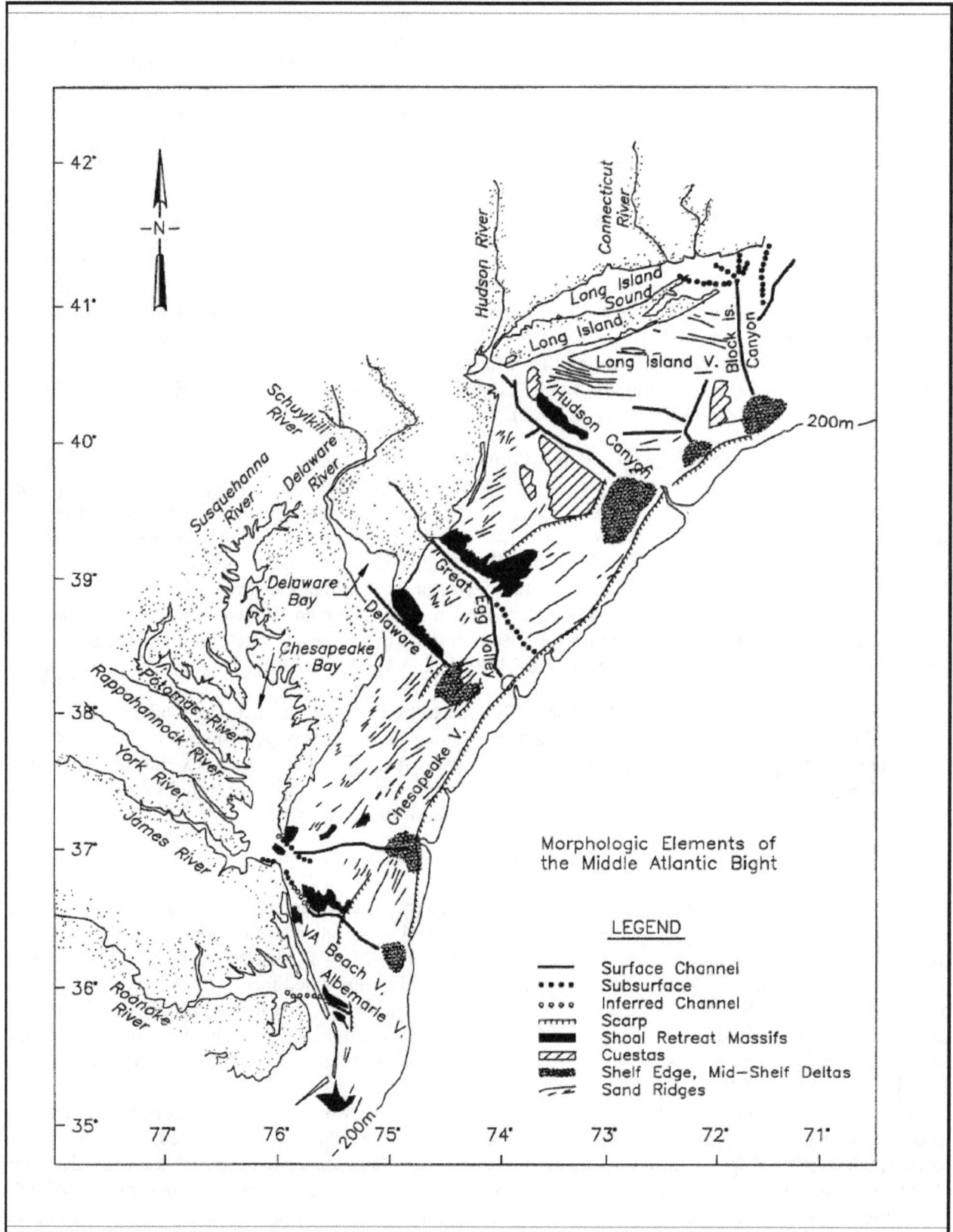

**Figure IV-2-36.** Morphology of the Middle Atlantic Bight (from Swift (1976)). Sand ridges close to shore may be suitable sources of sand for beach renourishment

## IV-2-14. References

**EM 1110-2-1502**
Coastal Littoral Transport

**Anders and Byrnes 1991**
Anders, F. J., and Byrnes, M. R. 1991. Accuracy of shoreline change rates as determined from maps and aerial photographs, *Shore and Beach*, Vol 51, No. 1, pp. 17-26.

**Baeteman 1994**
Baeteman, C. 1994. Subsidence in coastal lowlands due to groundwater withdrawal: the geological approach, "Coastal Hazards, Perception, Susceptibility and Mitigation," Finkl, C. W., Jr., (ed.), *Journal of Coastal Research* Special Issue No. 12, pp 61-75.

**Bagnold 1941**
Bagnold, R. A. 1941. *The Physics of Windblown Sand and Desert Dunes,* Methuen, London, UK.

**Bahr and Lanier 1981**
Bahr, L. M., and Lanier, W. P. 1981. "The Ecology of Intertidal Oyster Reefs of the South Atlantic Coast: A Community Profile," Report FWS/OBS-81/15, U.S. Fish and Wildlife Service, Office of Biological Services, Washington, DC.

**Ballard 1983**
Ballard, R. D. 1983. *Exploring our Living Planet*, National Geographic Society, Washington, DC.

**Bates and Jackson 1984**
Bates, R. L., and Jackson, J. A. 1984. *Dictionary of Geologic Terms,* 3rd ed., Anchor Press/Doubleday, Garden City, NY.

**Bowes 1989**
Bowes, M. A. 1989. "Review of the Geomorphic Diversity of the Great Lakes Shore Zone in Canada," Heritage Resources Centre, University of Waterloo, Waterloo, Canada.

**Boyd, Dalrymple, and Zaitlin 1992**
Boyd, R., Dalrymple, R., and Zaitlin, B. A. 1992. "Classification of Elastic Coastal Depositional Environments," *Sedimentary Geology*, Vol 80, pp 139-150.

**Bruun 1962**
Bruun, P. M. 1962. "Sea-level Rise as a Cause of Shore Erosion," *Proceedings, Journal of the Waterways and Harbor Division,* American Society of Civil Engineers, Vol 88, No. WW1, pp 117-130.

**Bullard 1962**
Bullard, F. M. 1962. *Volcanoes in History, in Theory, in Eruption*, University of Texas Press, Austin, TX.

**Carter 1988**
Carter, R. W. G. 1988. *Coastal Environments: An Introduction to the Physical, Ecological, and Cultural Systems of Coastlines,* Academic Press, London, UK.

**Carter, Curtis, and Sheehy-Skeffington 1992**
Carter, R. W. G., Curtis, T. G. F., and Sheehy-Skeffington, M. J. 1992. *Coastal Dunes, Geomorphology, Ecology, and Management for Conservation*, A. A. Balkema, Rotterdam, The Netherlands.

**Carter, Williams, Fuller, and Meisburger 1982**
Carter, C. H., Williams, S. J., Fuller, J. A., and Meisburger, E. P. 1982. "Regional Geology of the Southern Lake Erie (Ohio) Bottom: A Seismic Reflection and Vibracore Study," Miscellaneous Report No 82-15, Coastal Engineering Research Center, U.S. Army Engineer Waterways Experiment Station, Vicksburg, MS.

**Chapman 1974**
Chapman, V. J. 1974. "Salt Marshes and Salt Deserts of the World," *Ecology of Halophytes*, R. J. Reimold and W. H. Queen, eds., Academic Press, NY, pp 3-19.

**Coastal Barriers Study Group 1988**
Coastal Barriers Study Group. 1988. Report to Congress: Coastal Barrier Resources System, Recommendations for Additions to or Deletions from the Coastal Barrier Resources System, (in 22 Volumes covering the United States and territories), U.S. Department of the Interior, Washington, DC.

**Coleman and Wright 1971**
Coleman, J. M., and Wright, L. D. 1971. "Analysis of Major River Systems and Their Deltas: Procedures and Rationale, with Two Examples," Technical Report No. 95, Coastal Studies Institute, Louisiana State University, Baton Rouge, LA.

**Cotton 1952**
Cotton, C. A. 1952. "Criteria for the Classification of Coasts," Abstract of papers, 17th Congress of International Geographers, Washington, DC, p 15.

**Cromwell 1971**
Cromwell. 1971. Barrier Coast Distribution: A World-wide Survey," Abstracts, Second Coastal and Shallow Water Research Conference, U.S. Office of Naval Research Geography Program, University Press, University of Southern California, Los Angeles, CA, p 50.

**Crowell, Leatherman, and Buckley 1991**
Crowell, M., Leatherman, S. P., and Buckley, M. K. 1991. "Historical Shoreline Change: Error Analysis and Mapping Accuracy," *Journal of Coastal Research*, Vol 7, No. 3, pp 839-852.

**Dalrymple, Silver, and Jackson 1973**
Dalrymple, G. B., Silver, E. I., and Jackson, E. D. 1973. "Origin of the Hawaiian Islands," *American Scientist*, Vol 61, No. 3, pp 294-308.

**Dalrymple, Zaitlin, and Boyd 1992**
Dalrymple, R. W., Zaitlin, B. A., and Boyd, R. 1992. "Estuarine Facies Models: Conceptual Basis and Stratigraphic Implications," *Journal of Sedimentary Petrology*, Vol 62, No. 6, pp 1130-1146.

**Dana 1849**
Dana, J. D. 1849. "Geology," *Report of the U.S. Exploring Expedition, 1838-1842*, C. Sherman, Philadelphia, PA.

**Davis 1896a**
Davis, W. M. 1896a. "Shoreline Topography," reprinted in: *Geographical Essays,* Dover Publications, New York, NY (1954 reprint).

**Davis 1985**
Davis, R. A., Jr., ed. 1985. *Coastal Sedimentary Environments,* 2nd ed., Springer-Verlag, New York, NY.

**de Blij and Muller 1993**
de Blij, H. J., and Muller, P. O. 1993. *Physical Geography of the Global Environment*, John Wiley & Sons, New York, NY.

**Deery and Howard 1977**
Deery, J. R., and Howard, J. D. 1977. "Origin and Character of Washover Fans on the Georgia Coast, USA," *Gulf Coast Association Geologic Society Transactions,* Vol 27, pp 259-271.

**Dillon 1970**
Dillon, W. P. 1970. "Submergence Effects on a Rhode Island Barrier and Lagoon and Inferences on Migration of Barriers," *Journal of Geology*, Vol 78, pp 94-106.

**Dingler, Reiss, and Plant 1993**
Dingler, J. R., Reiss, T. E., and Plant, N. G. 1993. "Erosional Patterns of the Isles Dernieres, Louisiana, in Relation to Meteorological Influences," *Journal of Coastal Research*, Vol 9, No. 1, pp 112-125.

**Dolan and Hayden 1983**
Dolan, R., and Hayden, B. 1983. "Patterns and Prediction of Shoreline Change," *CRC Handbook of Coastal Processes,* P. D. Komar, ed., CRC Press, Boca Raton, FL, pp 123-165.

**Duane and Meisburger 1969**
Duane, D. B., and Meisburger, E. P. 1969. "Geomorphology and Sediments of the Nearshore Continental Shelf, Miami to Palm Beach, Florida," Technical Memorandum No. 29, Coastal Engineering Research Center, USAE Waterways Experiment Station, Vicksburg, MS.

**Duane, Field, Meisburger, Swift, and Williams 1972**
Duane, D. B., Field, M. E., Meisburger, E. P., Swift, D. J. P., and Williams, S. J. 1972. "Linear Shoals on the Atlantic Inner Continental Shelf, Florida to Long Island," *Shelf Sediment Transport*, D. J. P. Swift, D. B. Duane, and O. H. Pilkey, eds., Dowd, Hutchinson, and Ross, Stroudsburg, PA.

**Dyer 1979**
Dyer, K. R., ed. 1979. *Estuarine Hydrography and Sedimentation, a Handbook*, Cambridge University Press, Cambridge, UK.

**Edwards and Frey 1977**
Edwards, J. M., and Frey, R. W. 1977. "Substrate Characteristics Within a Holocene Salt Marsh, Sapelo Island, Georgia," *Senckenberg. Marit.,* Vol 9, pp 215-259.

**Emery 1967**
Emery, K. O. 1967. "Estuaries and Lagoons in Relation to Continental Shelves," *Estuaries*, Publication 83, G. H. Lauff, ed., American Association for the Advancement of Science, Washington, DC, pp 9-11.

**Emery 1968**
Emery, K. O. 1968. "Relict Sediments on Continental Shelves of the World," *American Association of Petroleum Geologists Bulletin,* Vol 52, No. 3, pp 445-464.

**Field 1979**
Field, M. E. 1979. "Sediments, Shallow Subbottom Structure, and Sand Resources of the Inner Continental Shelf, Central Delmarva Peninsula," Technical Paper No. 79-2, Coastal Engineering Research Center, U.S. Army Engineer Waterways Experiment Station, Vicksburg, MS.

**Field and Duane 1974**
Field, M. E., and Duane, D. B. 1974. "Geomorphology and Sediments of the Inner Continental Shelf, Cape Canaveral, Florida," Technical Memorandum No. 42, Coastal Engineering Research Center, USAE Waterways Experiment Station, Vicksburg, MS.

**Fisher and Dolan 1977**
Fisher, J. S., and Dolan, R. 1977. "Beach Processes and Coastal Hydrodynamics," *Benchmark Papers in Geology,* Vol 39, Dowden, Hutchinson & Ross, Stroudsburg, PA.

**FitzGerald and Rosen 1987**
FitzGerald, D. M., and Rosen, P. S., eds. 1987. *Glaciated Coasts,* Academic Press, San Diego, CA.

**Flint 1971**
Flint, R. F. 1971. *Glacial and Quaternary Geology,* John Wiley and Sons, New York, NY.

**Frey and Basan 1985**
Frey, R. W., and Basan, P. B. 1985. Coastal Salt Marshes, *Coastal Sedimentary Environments,* 2nd ed., R. A. Davis, Jr., ed., Springer-Verlag, New York, pp 225-303.

**Fuller and Meisburger 1982**
Fuller, J. A., and Meisburger, E. P. 1982. "A Lightweight Pneumatic Coring Device: Design and Field Test," Miscellaneous Report No. 82-8, Coastal Engineering Research Center, U.S. Army Engineer Waterways Experiment Station, Vicksburg, MS.

**Giese 1988**
Giese, G. S. 1988. "Cyclical Behavior of the Tidal Inlet at Nauset Beach, Chatham, Massachusetts," *Hydrodynamics and Sediment Dynamics of Tidal Inlets,* Lecture Notes on Coastal and Estuarine Studies, Vol 29, D. G. Aubrey and L. Weishar, eds., Springer-Verlag, New York, NY, pp 269-283.

**Gilbert 1885**
Gilbert, G. K. 1885. 5th Annual Report, U.S. Geological Survey, Washington, DC, pp 87-88 (reprinted in Schwartz (1973)).

**Goldsmith 1985**
Goldsmith, V. 1985. "Coastal Dunes," Coastal Sedimentary Environments, 2nd ed., R. A. Davis, Jr., ed., Springer-Verlag, New York, NY, pp 303-378.

**Gorman, Morang, and Larson 1998**
Gorman, L. T., Morang, A., and Larson, R. L. 1998. Monitoring the Coastal Environment; Part IV: Mapping, Shoreline Change, and Bathymetric Analysis, *Journal of Coastal Research,* Vol 14, No. 1, pp 61-92.

**Guilcher 1965**
Guilcher, A. 1965. "Drumlin and Spit Structures in the Kenmare River, Southwest Ireland," *Irish Geography*, Vol 5, No. 2, pp 7-19.

**Gulliver 1899**
Gulliver, F. P. 1899. "Shoreline Topography," *Proceedings of the American Academy of Arts and Science*, Vol 34, pp 149-258.

**Hands 1983**
Hands, E. B. 1983. "The Great Lakes as a Test Model for Profile Response to Sea Level Changes," Chapter 8 in *Handbook of Coastal Processes and Erosion*, P. D. Komar, ed., CRC Press, Inc., Boca Raton, FL. (Reprinted in Miscellaneous Paper CERC-84-14, Coastal Engineering Research Center, U.S. Army Engineer Waterways Experiment Station, Vicksburg, MS.)

**Hathaway 1972**
Hathaway, J. C. 1972. "Regional Clay Mineral Facies in Estuaries and Continental Margin of the United States and East Coast," *Environmental Framework of Coastal Plain Estuaries*, B. W. Nelson, ed., Geologic Society of America Memoir 133, Boulder, CO, pp 293-316.

**Hayes 1980**
Hayes, M. O. 1980. "General Morphology and Sediment Patterns in Tidal Inlets," *Sedimentary Geology*, Vol 26, pp 139-156.

**Herdendorf 1988**
Herdendorf, C. E. 1988. "Classification of Geologic Features in Great Lakes Nearshore and Coastal Areas," Committee report prepared for: Protecting Great Lakes Nearshore and Coastal Diversity Project, International Joint Commission, Windsor, Ontario.

**Hotta, Kraus, and Horikawa 1991**
Hotta, S., Kraus, N. C., and Horikawa, K. 1991. "Functioning of Multi-row Sand Fences in Forming Foredunes," *Proceedings of Coastal Sediments '91*, American Society of Civil Engineers, New York, pp 261-275.

**Howard and Frey 1977**
Howard, J. D., and Frey, R. W. 1977. "Characteristic Physical and Biogenic Sedimentary Structures in Georgia Estuaries," *American Association of Petroleum Geologist Bulletin*, Vol 57, pp 1169-1184.

**Hoyt 1967**
Hoyt, J. H. 1967. "Barrier Island Formation," *Bulletin of the Geological Society of America*, Vol 78, pp 1125-1136.

**Hume and Herdendorf 1988**
Hume, T. M., and Herdendorf, C. E. 1988. "A Geomorphic Classification of Estuaries and Its Application to Coastal Resource Management - a New Zealand Example," *Journal of Ocean and Shoreline Management*, Vol 11, pp 249-274.

**Inman and Nordstrom 1971**
Inman, D. L., and Nordstrom, C. E. 1971. "On the Tectonic and Morphological Classification of Coasts," *Journal of Geology*, Vol 79, pp 1-21.

**Johnson 1919**

Johnson, D. 1919. *Shore Processes and Shoreline Development,* John Wiley & Sons, New York, NY.

**Kerwin and Pedigo 1971**

Kerwin, J. A., and Pedigo, R. A. 1971. "Synecology of a Virginia Salt Marsh," *Chesapeake Science,* Vol 12, pp 125-130.

**King 1972**

King, C. A. M. 1972. *Beaches and Coasts,* 2nd ed., Edward Arnold, London, UK.

**Knutson 1976**

Knutson, P. L. 1976. "Summary of CERC Research on Uses of Vegetation for Erosion Control," *Proceedings of Great Lakes Vegetation Workshop,* Great Lakes Basin Commission and USDA Soil Conservation Service, pp 31-36.

**Knutson 1978**

Knutson, P. L. 1978. "Planting Guidelines for Dune Creation and Stabilization," *Proceedings of Symposium on Technical, Environmental, Socioeconomic and Regulatory Aspects of Coastal Zone Planning and Management,* American Society of Civil Engineers, Vol 2, pp 762-779.

**Kolb and van Lopik 1966**

Kolb, C. R., and van Lopik, J. R. 1966. "Depositional Environments of the Mississippi River Deltaic Plain-Southeastern Louisiana," In: *Deltas in Their Geologic Framework,* Houston Geological Society, Houston, TX, pp 17-61.

**Komar 1976**

Komar, P. D. 1976. *Beach Processes and Sedimentation,* Prentice-Hall, Englewood Cliffs, NJ.

**Komar 1983**

Komar, P. D. 1983. "Computer Models of Shoreline Changes," *CRC Handbook of Coastal Processes and Erosion,* P. D. Komar, ed., CRC Press, Inc., Boca Raton, FL, pp 205-216.

**Lauff 1967**

Lauff, G. H., ed. 1967. *Estuaries,* American Association for the Advancement of Science, Pub. No. 83, Washington, DC.

**Leatherman 1979**

Leatherman, S. P., ed. 1979. *Barrier Islands from the Gulf of St. Lawrence to the Gulf of Mexico,* Academic Press, New York, NY.

**McBride and Moslow 1991**

McBride, R. A., and Moslow, T. F. 1991. "Origin, Evolution, and Distribution of Shoreface Sand Ridges, Atlantic Inner Shelf, U.S.A.," *Marine Geology,* Vol 97, pp 57-85.

**Meisburger 1972**

Meisburger, E. P. 1972. "Geomorphology and Sediments of the Chesapeake Bay Entrance," Technical Memorandum No. 38, Coastal Engineering Research Center, U.S. Army Engineer Waterways Experiment Station, Vicksburg, MS.

**Meisburger 1976**
Meisburger, E. P. 1976. "Geomorphology and Sediments of Western Massachusetts Bay," Technical Paper No. 76-3, Coastal Engineering Research Center, U.S. Army Engineer Waterways Experiment Station, Vicksburg, MS.

**Meisburger 1977**
Meisburger, E. P. 1977. "Sand Resources on the Inner Continental Shelf of the Cape Fear Region," Miscellaneous Report No. 77-11, Coastal Engineering Research Center, U.S. Army Engineer Waterways Experiment Station, Vicksburg, MS.

**Meisburger 1979**
Meisburger, E. P. 1979. "Reconnaissance Geology of the Inner Continental Shelf, Cape Fear Region, North Carolina," Technical Report No. 79-3, Coastal Engineering Research Center, U.S. Army Engineer Waterways Experiment Station, Vicksburg, MS.

**Meisburger and Field 1975**
Meisburger, E. P., and Field, M. E. 1975. "Geomorphology, Shallow Structure, and Sediments of the Florida Inner Continental Shelf, Cape Canaveral to Georgia," Technical Memorandum No. 54, Coastal Engineering Research Center, U.S. Army Engineer Waterways Experiment Station, Vicksburg, MS.

**Meisburger and Williams 1980**
Meisburger, E. P., and Williams, S. J. 1980. "Sand Resources on the Inner Continental Shelf of the Cape May Region, New Jersey," Miscellaneous Report 80-4, Coastal Engineering Research Center, U.S. Army Engineer Waterways Experiment Station, Vicksburg, MS.

**Meisburger and Williams 1981**
Meisburger, E. P., and Williams, S. J. 1981. "Use of Vibratory Coring Samplers for Sediment Surveys," Coastal Engineering Technical Aid No. 81-9, Coastal Engineering Research Center, U.S. Army Engineer Waterways Experiment Station, Vicksburg, MS.

**Meisburger and Williams 1982**
Meisburger, E. P., and Williams, S. J. 1982. "Sand Resources on the Inner Continental Shelf Off the Central New Jersey Coast," Miscellaneous Report No. 82-10, Coastal Engineering Research Center, U.S. Army Engineer Waterways Experiment Station, Vicksburg, MS.

**Meisburger, Williams, and Prins 1979**
Meisburger, E. P., Williams, S. J., and Prins, D. A. 1979. "Sand Resources of Southeastern Lake Michigan," Miscellaneous Report No. 79-3, Coastal Engineering Research Center, U.S. Army Engineer Waterways Experiment Station, Vicksburg, MS.

**Metha 1986**
Metha, A. J., ed. 1986. *Estuarine Cohesive Sediment Dynamics*, Lecture Notes on Coastal and Estuarine Studies, Vol 14, Springer-Verlag, New York, NY.

**Moore and Clague 1992**
Moore, J. G., and Clague, D. A. 1992. "Volcano Growth and Evolution of the Island of Hawaii," *Geological Society of America Bulletin*, Vol 104, No. 11, pp 1471-1484.

**Nelson 1972**
Nelson, B. W., ed. 1972. *Environmental Framework of Coastal Plain Estuaries*, Geological Society of America Memoir 133, Boulder, CO.

**Nichols 1968**
Nichols, R. L. 1968. "Coastal Geomorphology, McMurdo Sound, Antarctica," *Journal of Glaciology,* Vol 51, pp 694-708.

**Nichols and Biggs 1985**
Nichols, M. M., and Biggs, R. B. 1985. "Estuaries," *Coastal Sedimentary Environments*, R. A. Davis, Jr., ed., 2nd ed., Springer-Verlag, New York, NY, pp 77-186.

**Nummedal 1983**
Nummedal, D. 1983. "Barrier Islands," *CRC Handbook of Coastal Processes and Erosion*, P. D. Komar, ed., CRC Press, Inc., Boca Raton, FL, pp 77-121.

**Otvos 1970**
Otvos, E. G., Jr. 1970. "Development and Migration of Barrier Islands, Northern Gulf of Mexico," *Bulletin of the Geological Society of America,* Vol 81, pp 241-246.

**Otvos 1981**
Otvos, E. G., Jr. 1981. "Barrier Island Formation Through Nearshore Aggradation - Stratigraphic and Field Evidence," *Marine Geology,* Vol 43, pp 195-243.

**Penland and Boyd 1981**
Penland, S., and Boyd, R. 1981. "Shoreline Changes on the Louisiana Barrier Coast," *Proceedings of the Oceans '81 Conference,* Boston, Massachusetts, September 16-18, pp 209-219.

**Pethick 1984**
Pethick, J. 1984. *An Introduction to Coastal Geomorphology,* Edward Arnold Publishers, London, UK.

**Pettijohn 1975**
Pettijohn, F. J. 1975. *Sedimentary Rocks,* Harper and Row, New York, NY.

**Pilkey and Field 1972**
Pilkey, O. H., and Field, M. E. 1972. "Onshore Transportation of Continental Shelf Sediment: Atlantic Southeastern United States," *Shelf Sediment Transport,* D. J. Swift, D. B. Duane, and O. H. Pilkey, eds., Dowden, Hutchinson, & Ross, Stroudsburg, PA.

**Press and Siever 1986**
Press, F., and Siever, R. 1986. *Earth,* 4th. ed., W. H. Freeman and Company, New York, NY.

**Prins 1980**
Prins, D. A. 1980. "Data Collection Methods for Sand Inventory-Type Surveys," Coastal Engineering Technical Aid No. 80-4, Coastal Engineering Research Center, U.S. Army Engineer Waterways Experiment Station, Vicksburg, MS.

**Pritchard 1967**
Pritchard, D. W. 1967. "What is an Estuary? Physical Viewpoint," *Estuaries*, Publication 83, G. H. Lauff, ed., American Association for the Advancement of Science, Washington, DC, pp 3-5.

**Reading 1986**
Reading, H. G., ed. 1986. *Sedimentary Environments and Facies,* 2nd ed., Blackwell Scientific Publications, Oxford, UK.

**Reineck and Singh 1980**
Reineck, H. E., and Singh, I. B. 1980. *Depositional Sedimentary Environments,* 2nd ed., Springer-Verlag, Berlin, Germany.

**Russell 1967**
Russell, R. 1967. Origin of estuaries, *Estuaries,* G. H. Lauff, (ed.), American Association for the Advancement of Science, Publication 83, Washington, D.C., pp 93-99.

**Schwartz 1971**
Schwartz, M. L. 1971. "The Multiple Casualty of Barrier Islands," *Journal of Geology,* Vol 79, pp 91-94.

**Schwartz 1973**
Schwartz, M. L., ed. 1973. *Barrier Islands,* Dowden, Hutchinson & Ross, Stroudsburg, PA.

**Schwartz 1982**
Schwartz, M. L., ed. 1982. *The Encyclopedia of Beaches and Coastal Environments,* Encyclopedia of Earth Sciences, Volume XV, Hutchinson Ross Publishing Company, Stroudsburg, PA.

**Shepard 1937**
Shepard, F. P. 1937. "Revised Classification of Marine Shorelines," *Journal of Geology,* Vol 45, pp 602-624.

**Shepard 1948**
Shepard, F. P. 1948. *Submarine Geology,* Harper & Row, New York, NY.

**Shepard 1973**
Shepard, F. P. 1973. *Submarine Geology,* 3rd ed., Harper & Row, New York, NY.

**Shepard 1976**
Shepard, F. P. 1976. "Coastal Classification and Changing Coastlines," *Geoscience and Man,* Vol 13, pp 53-64.

**Shepard 1977**
Shepard, F. P. 1977. *Geological Oceanography,* Crane, Russak & Co., New York, NY.

**Shepard and Wanless 1971**
Shepard, F. P., and Wanless, H. R. 1971. *Our Changing Coastlines,* McGraw-Hill Book Company, New York, NY.

**Smith 1954**
Smith, H. T. U. 1954. "Coastal Dunes," *Proceedings of the Coastal Geography Conference,* Office of Naval Research, Department of the Navy, Washington, DC, pp 51-56.

**Spalding, Ravilious, and Green 2001**
Spalding, M. D., Ravilious, C., and Green, E. P. 2001. *World Atlas of Coral Reefs,* The University of California Press, Berkeley, CA.

**Spurgeon 1992**
Spurgeon, J. P. G. 1992. "The Economic Valuation of Coral Reefs," *Marine Pollution Bulletin*, Vol 24, No. II, pp 529-536.

**Stewart and Pope 1992**
Stewart, C. J., and Pope, J. 1992. "Erosion Processes Task Group Report," Open file report prepared for the International Joint Commission Water Level Reference Study, International Joint Commission, Canada and U.S.A.

**Stoddard 1969**
Stoddard, D. R. 1969. "Ecology and Morphology of Recent Coral Reefs," *Biological Reviews*, Vol 44, pp 433-498.

**Strahler 1971**
Strahler, A. N. 1971. *The Earth Sciences*, 2nd ed., Harper & Row Publishers, New York, NY.

**Suess 1888**
Suess, E. 1888. *The Faces of the Earth,* Vol 2 (English translation by H. B. Sollas in 1906), Oxford University Press, London, UK (in 5 vols).

**Suhayda 1984**
Suhayda, J. N. 1984. "Interaction Between Surface Waves and Muddy Bottom Sediments," *Estuarine Cohesive Sediment Dynamics,* Lecture Notes on Coastal and Estuarine Studies, A. J. Mehta, ed., Springer-Verlag, Berlin, pp 401-428.

**Suter and Berryhill 1985**
Suter, J. R., and Berryhill, H. L., Jr. 1985. "Late Quaternary Shelf-Margin Deltas, Northwest Gulf of Mexico," *Bulletin of the American Association of Petroleum Geologists,* Vol 69, No. 1, pp 77-91.

**Swift 1976**
Swift, D. J. P. 1976. "Coastal Sedimentation," *Marine Sediment Transport and Environmental Management*, D. J. Stanley and D. J. P. Swift, eds., John Wiley and Sons, New York, NY, pp 255-310.

**Tilling, Heliker, and Wright 1987**
Tilling, R. I., Heliker, C., and Wright, T. L. 1987. "Eruptions of Hawaiian Volcanoes: Past, Present, and Future," U.S. Geological Survey, Denver, CO.

**Trenhaile 1987**
Trenhaile, A. S. 1987. *The Geomorphology of Rock Coasts*, Clarendon Press, Oxford, UK.

**U.S. Army Engineer Waterways Experiment Station 1992**
U.S. Army Engineer Waterways Experiment Station. 1992. "The Wetlands Research Program Notebook," Technical Notes, U.S. Army Engineer Waterways Experiment Station, Vicksburg, MS.

**Valentin 1952**
Valentin, H. 1952. *Die Kusten der Erde,* Petermanns Geog. Mitt. Erg. 246, Justus Perthes Gotha, Berlin, Germany.

**Van de Kreeke 1986**
Van de Kreeke, J., ed. 1986. *Physics of Shallow Estuaries and Bays*, Lecture Notes on Coastal and Estuarine Studies, Vol 16, Springer-Verlag, New York, NY.

**Waisel 1972**
Waisel, Y. 1972. *Biology of Halophytes*, Academic Press, NY.

**Williams 1976**
Williams, S. J. 1976. "Geomorphology, Shallow Subbottom Structure, and Sediments of the Atlantic Inner Continental Shelf Off Long Island, New York," Technical Paper No. 76-2, Coastal Engineering Research Center, U.S. Army Engineer Waterways Experiment Station, Vicksburg, MS.

**Williams 1981**
Williams, S. J. 1981. "Sand Resources and Geological Character of Long Island Sound," Technical Paper No. 81-3, Coastal Engineering Research Center, U.S. Army Engineer Waterways Experiment Station, Vicksburg, MS.

**Williams and Duane 1974**
Williams, S. J., and Duane, D. B. 1974. "Geomorphology and Sediments of the Inner New York Bight Continental Shelf," Technical Memorandum No. 45, Coastal Engineering Research Center, U.S. Army Engineer Waterways Experiment Station, Vicksburg, MS.

**Williams and Meisburger 1982**
Williams, S. J., and Meisburger, E. P. 1982. "Geological Character and Mineral Resources of South Central Lake Erie," Miscellaneous Report No. 82-9, Coastal Engineering Research Center, U.S. Army Engineer Waterways Experiment Station, Vicksburg, MS.

**Williams, Carter, Meisburger, and Fuller 1980**
Williams, S. J., Carter, C. H., Meisburger, E. P., and Fuller, J. A. 1980. "Sand Resources of Southern Lake Erie, Conneaut to Toledo, Ohio - a Seismic Reflection and Vibracore Study," Miscellaneous Report No. 80-10, Coastal Engineering Research Center, U.S. Army Engineer Waterways Experiment Station, Vicksburg, MS.

**Williams, Prins, and Meisburger 1979**
Williams, S. J., Prins, D. A., and Meisburger, E. P. 1979. "Sediment Distribution Sand Resources, and Geologic Character of the Inner Continental Shelf Off Galveston County Texas," Miscellaneous Report No. 79-4, Coastal Engineering Research Center, U.S. Army Engineer Waterways Experiment Station, Vicksburg, MS.

**Wood and Kienle 1990**
Wood, C. A., and Kienle, J., eds. 1990. *Volcanoes of North America: United States and Canada*, Cambridge University Press, Cambridge, UK.

**Woodhouse 1978**
Woodhouse, W. W., Jr. 1978. "Dune Building and Stabilization with Vegetation," SR-3, Coastal Engineering Research Center, U.S. Army Engineer Waterways Experiment Station, Vicksburg, MS.

**Young and Hale 1998**
Young, C., and Hale, L. 1998. "Coastal Management: Insurance for the Coastal Zone," *Maritimes*, Vol 40m, No. 1, pp.17-19.

## IV-2-15. Acknowledgments

Authors of Chapter IV-2, "Coastal Classification and Morphology:"

Andrew Morang, Ph.D., Coastal and Hydraulics Laboratory (CHL), Engineer Research and Development
Center, Vicksburg, Mississippi.
Laurel T. Gorman, Information Technology Laboratory, Engineer Research and Development Center,
Vicksburg, Mississippi.
David B. King, Ph.D., CHL.
Edward Meisburger, CHL (retired).

Reviewers:

Stephan A. Chesser, U.S. Army Engineer District, Portland, Portland, Oregon.
Ronald L. Erickson, U.S. Army Engineer District, Detroit, Detroit, Michigan.
Joan Pope, CHL.
John F. C. Sanda, Headquarters, U.S. Army Corps of Engineers, Washington, DC., (retired).
Orson P. Smith, Ph.D., U.S. Army Engineer District, Alaska, Anchorage, Alaska, (retired).

## Table of Contents

## List of Tables

# List of Figures

**Chapter IV-3
Coastal Morphodynamics**

## IV-3-1. Introduction

*a.* This chapter discusses the morphodynamics of three coastal environments: deltas, inlets, and sandy shores. The divisions are somewhat arbitrary because, in many circumstances, the environments are found together in limited areas. For example, within a major river delta like the Mississippi, there are sandy beaches, bays where cohesive sediments accumulate, and inlets which funnel water in and out of the bays. Coastal features and environments are also not constant over time. For example, as we discussed in Part IV-2, estuaries, deltas, and beach ridge shores are elements of a landform continuum that extends over time. Which particular environment or shore type is found at any one time depends on sea level rise, sediment supply, wave and tide energy, underlying geology, climate, rainfall, runoff, and biological productivity.

*b.* Based on the fact that physical conditions along the coast are constantly changing, it can be argued that there is no such thing as an "equilibrium" state for any coastal form. This is true not only for shoreface profiles but also for deltas, which continue to shift over time in response to varying wave and meteorologic conditions. In addition, man continues to profoundly influence the coastal environment throughout the world, changing natural patterns of runoff and littoral sediment supply and constantly rebuilding and modifying engineering works. This is true even along undeveloped coastlines because of environmental damage such as deforestation, which causes drastic erosion and increased sediment load in rivers. The reader is urged to remember that coastal landforms are the result of the interactions of a myriad of physical processes, man-made influences, global tectonics, local underlying geology, and biology.

*c.* Cohesive shores, another one of the primary geologic terrains found around the world, have been discussed in Part III-5, "Erosion, Transport, and Deposition of Cohesive Shores."

## IV-3-2. Introduction to Bed Forms

*a. Introduction.* When sediment is moved by flowing water, the individual grains are usually organized into morphological elements called *bed forms*. These occur in a baffling variety of shapes and scales. Some bed forms are stable only between certain values of flow strength. Often, small bed forms (ripples) are found superimposed on larger forms (dunes), suggesting that the flow field may vary dramatically over time. Bed forms may move in the same direction as the current flow, may move against the current (antidunes), or may not move at all except under specific circumstances. The study of bed form shape and size is of great value because it can assist in making quantitative estimates of the strength of currents in modern and ancient sediments (Harms 1969, Jopling 1966). Bed form orientations are indicators of flow pathways. This introduction to a complex subject is by necessity greatly condensed. For details on interpretation of surface structures and sediment laminae, readers are referred to textbooks on sedimentology such as Allen (1968, 1984, 1985); Komar (1998); Leeder (1982); Lewis (1984); Middleton (1965); Middleton and Southard (1984); and Reineck and Singh (1980).

*b. Environments.* In nature, bed forms are found in three environments with greatly differing characteristics:

(1) Rivers - unidirectional and channelized; large variety of grain sizes.

(2) Sandy coasts and bays - semi-channelized, unsteady, reversing (tidal) flows.

(3) Continental shelves - deep, unchannelized; dominated by geostrophic flows, storms, tidal currents, wave-generated currents.

*c. Classification.* Because of the diverse natural settings and the differing disciplines of researchers who have studied sedimentology, the classification and nomenclature of bed forms have been confusing and contradictory. The following classification scheme, proposed by the Society for Sedimentary Geology (SEPM) Bed forms and Bedding Structures Research Group in 1987 (Ashley 1990) is suitable for all subaqueous bed forms:

(1) Ripples. These are small bed forms with crest-to-crest spacing less than about 0.6 m and height less than about 0.03 m. It is generally agreed that ripples occur as assemblages of individuals similar in shape and scale. On the basis of crestline trace, Allen (1968) distinguished five basic patterns of ripples: straight, sinuous, catenary, linguoid, and lunate (Figure IV-3-1). The straight and sinuous forms may be symmetrical in cross section if subject to primarily oscillatory motion (waves) or may be asymmetrical if influenced by unidirectional flow (rivers or tidal currents). Ripples form a population distinct from larger-scale dunes, although the two forms share a similar geometry (Figure IV-3-2). The division between the two populations is caused by the interaction of ripple morphology and bed, and possibly shear stress. At low shear stresses, ripples are formed. As shear stress increases above a certain threshold, a "jump" in behavior occurs, resulting in the appearance of the larger dunes (Allen 1968).

(2) Dunes. Dunes are flow-transverse bed forms with spacings from under 1 m to over 1,000 m that develop on a sediment bed under unidirectional currents. These large bed forms are ubiquitous in sandy environments where water depths are greater than about 1 m, sand size coarser than 0.15 mm (very fine sand), and current velocities are greater than about 0.4 m/sec. In nature, these flow-transverse forms exist as a continuum of sizes without natural breaks or groupings (Ashley 1990). For this reason, "dune" replaces terms such as megaripple or sand wave, which were defined on the basis of arbitrary or perceived size distributions. Unfortunately, the term "sand wave" is still used in the literature, often with only the vaguest indication of what size feature is being described. For descriptive purposes, dunes can be subdivided as small (0.6- to 5-m wavelength), medium (5-10 m), large (10-100 m), and very large (> 100 m). In addition, the variation in pattern across the flow must be specified. If the flow pattern is relatively unchanged perpendicular to its overall direction and there are no eddies or vortices, the resulting bed form will be straight-crested and can be termed two-dimensional (Figures IV-3-3a and IV-3-4). If the flow structure varies significantly across the predominant direction and vortices capable of scouring the bed are present, a three-dimensional bed form is produced (Figure IV-3-5).

(3) Plane beds. A plane bed is a horizontal bed without elevations or depressions larger than the maximum size of the exposed sediment. The resistance to flow is small, resulting from grain roughness, which is a function of grain size. Plane beds occur under two hydraulic conditions:

(a) The transition zone between the region of no movement and the initiation of dunes (Figure IV-3-6).

(b) The transition zone between ripples and antidunes, at mean flow velocities between about 1 and 2 m/sec (Figure IV-3-6).

(4) Antidunes. Antidunes are bed forms that are in phase with water-surface gravity waves. Height and wavelength of these bedforms depend on the scale of the system and characteristics of the fluid and bed material (Reineck and Singh 1980). Trains of antidunes gradually build up from a plane bed as water velocity increases. As the antidunes increase in size, the water surface changes from planar to wave-like. The water waves may grow until they are unstable and break. As the sediment antidunes grow, they may

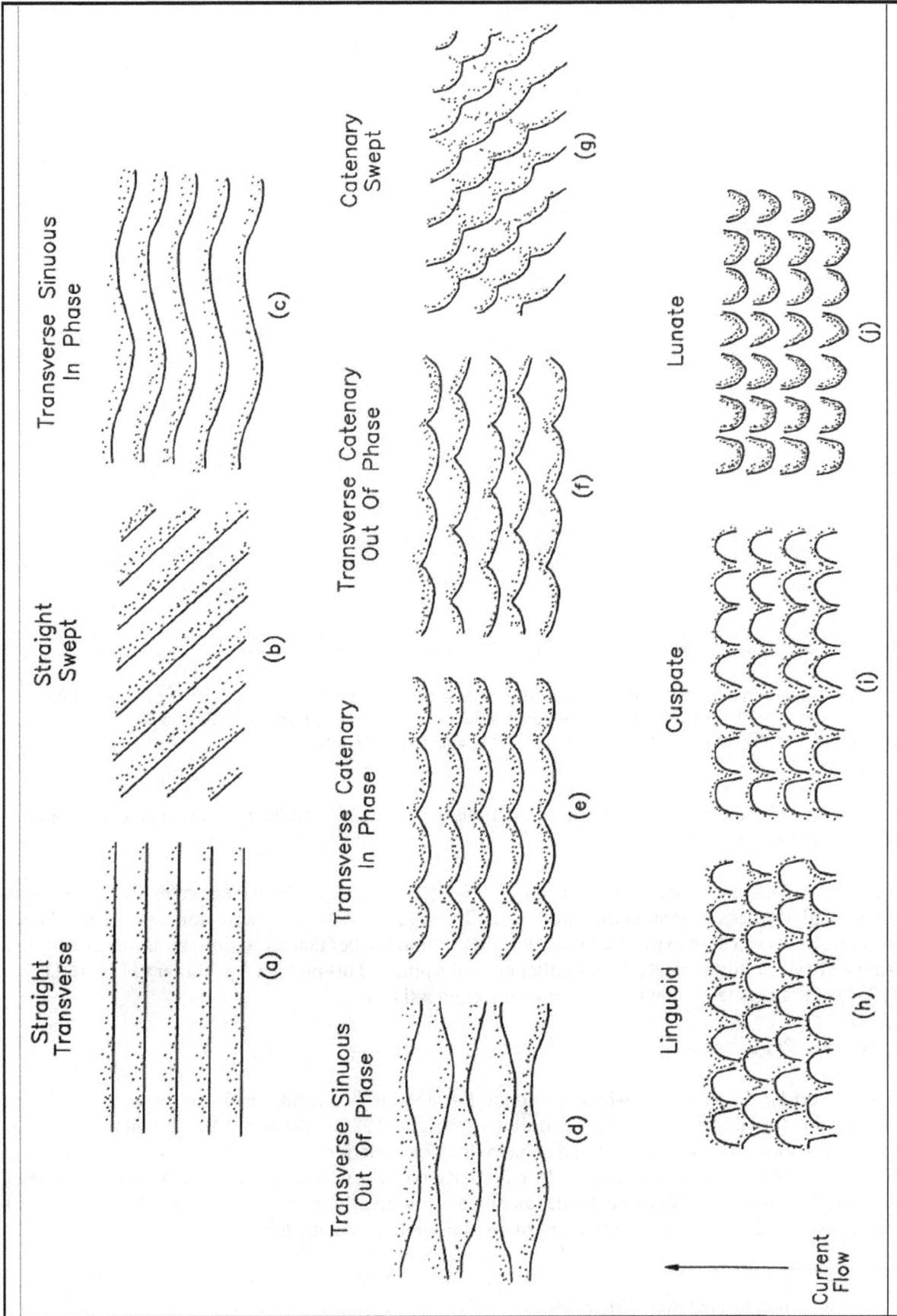

Straight
Transverse

(a)

Straight
Swept

(b)

Transverse Sinuous
In Phase

(c)

Transverse Sinuous
Out Of Phase

(d)

Transverse Catenary
In Phase

(e)

Transverse Catenary
Out Of Phase

(f)

Catenary
Swept

(g)

Linguoid

(h)

Cuspate

(i)

Lunate

(j)

Current
Flow

Figure IV-3-1. Sediment ripples. Water flow is from bottom to top, and lee sides and spurs are stippled (modified from Allen (1968))

**Figure IV-3-2. Ripples in a runnel, now exposed during low tide. Flow was from upper left to lower right. Complicated pattern is best classified as transverse sinuous out of phase (type d) in Figure IV-3-1. Photographed at Bon Secour Wildlife Refuge, near Gulf Shores, Alabama**

migrate upstream or downstream, or may remain stationary (the name "antidune" is based on early observations of upstream migration).

*d. Velocity - grain size relationships.* Figure IV-3-6 (from Ashley (1990)) illustrates the zones where ripples, dunes, planar beds, and antidunes are found. The figure summarizes laboratory studies conducted by various researchers. These experiments support the common belief that large flow-transverse bed forms (dunes) are a distinct entity separate from smaller current ripples. This plot is very similar to Figure 11.4 in Graf's (1984) hydraulics text, although Graf uses different axis units.

## IV-3-3. Deltaic Processes[1]

*a. Introduction.* River deltas, which are found throughout the world, result from the interaction of fluvial and marine (or lacustrine) forces. According to Wright (1985), "deltas are defined more broadly as coastal accumulations, both subaqueous and subaerial, of river-derived sediments adjacent to, or in close proximity to, the source stream, including the deposits that have been secondarily molded by waves, currents, or tides." The processes that control delta development vary greatly in intensity around the world. As a result, delta-plain landforms span the spectrum of coastal features and include:

---

[1] Material in this section adapted from Wright (1985).

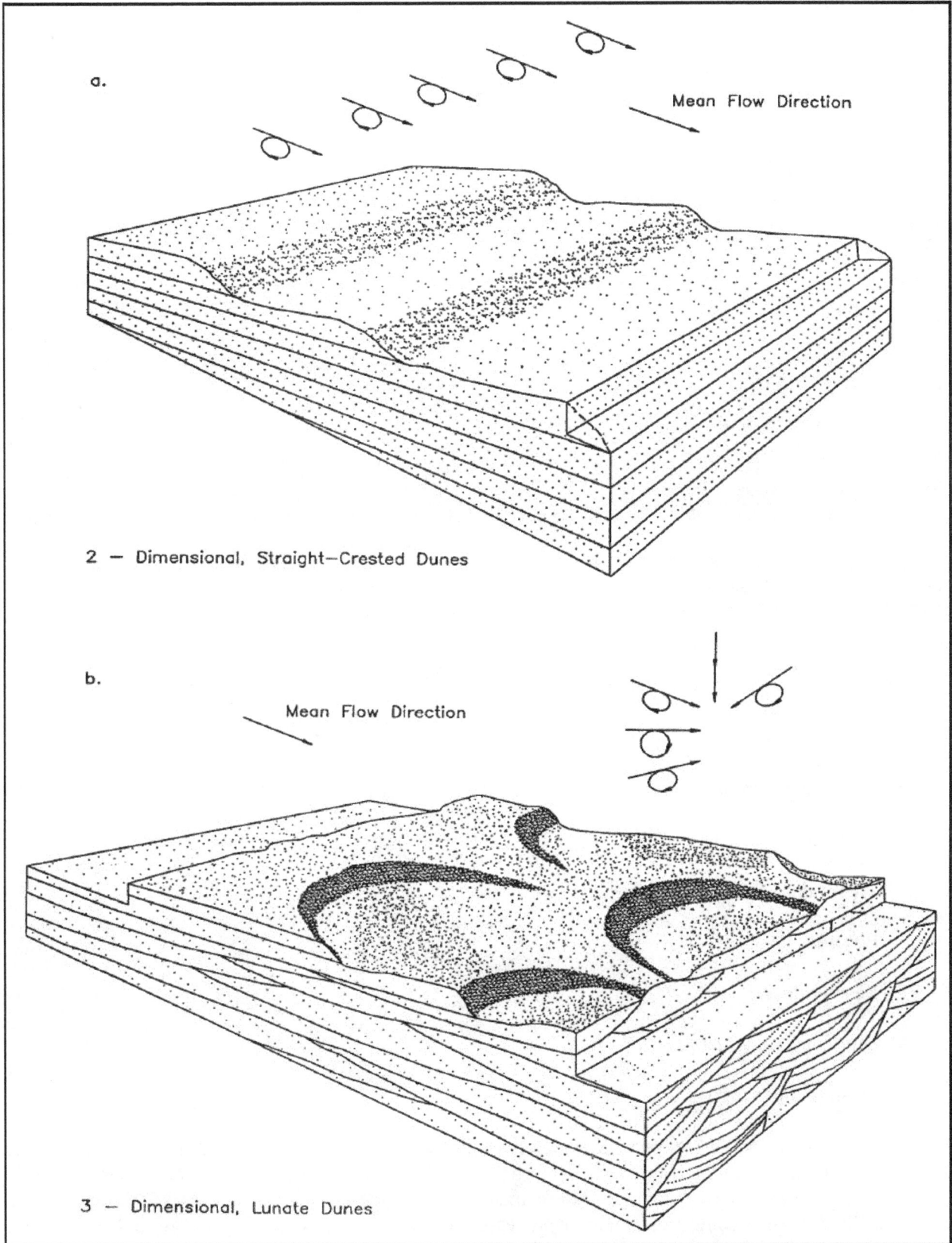

Figure IV-3-3. Two-dimensional and three-dimensional dunes. Vortices and flow patterns are shown by arrows above dunes. Adapted from Reineck and Singh (1980)

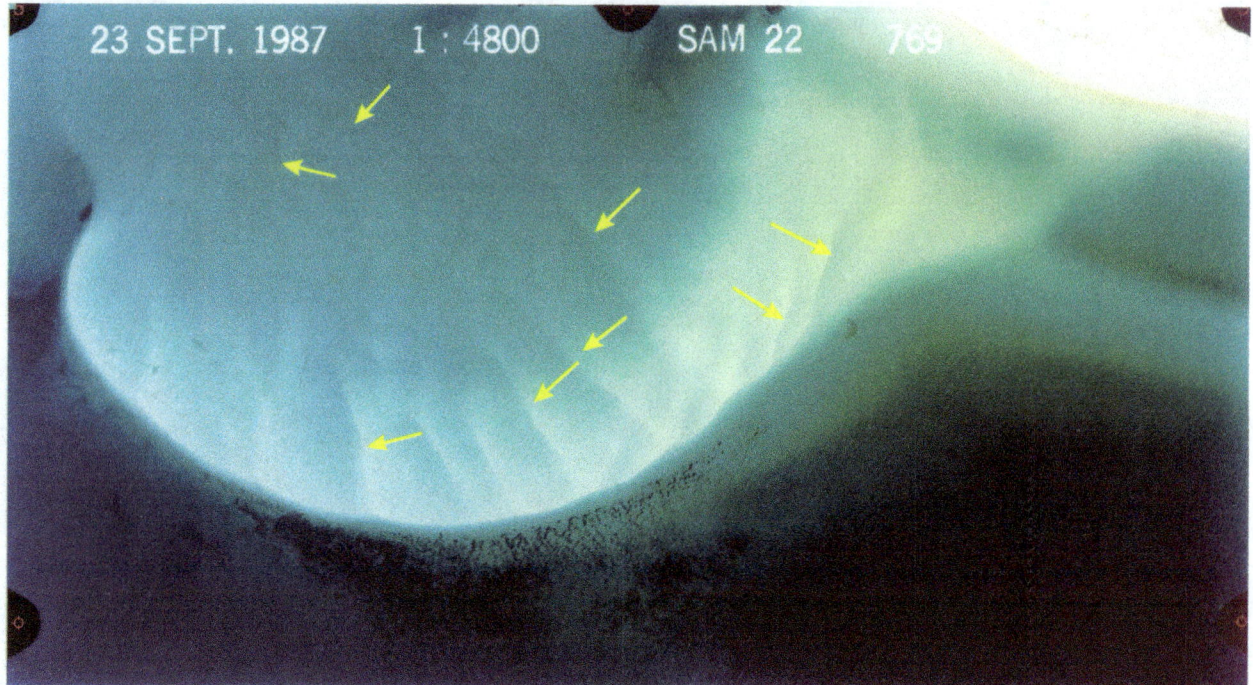

Figure IV-3-4. Ebb shoal, East Pass, Florida (23 Sep 1987). The clear water reveals two-dimensional dunes on the ebb shield. Water depth at the edge of the shoal is 3-4 m. North is to the top of the image. Distance from left to right is about 1 km (photograph of USAED, Mobile)

(1) Distributary channels.

(2) River-mouth bars.

(3) Interdistributary bays.

(4) Tidal flats.

(5) Tidal ridges.

(6) Beaches.

(7) Beach ridges.

(8) Dunes and dune fields.

(9) Swamps and marshes.

Despite the pronounced variety of worldwide environments where deltas are found, all actively-forming deltas have at least one common attribute: a river supplies clastic sediments to the coast and inner shelf more rapidly than marine processes can remove these materials. Whether a river is sufficiently large to transport enough sediment to overcome erosive marine processes depends upon the climate, geology, and nature of the drainage basin, and, most important, the overall size of the basin. The following paragraphs discuss delta classification, riverine flow, sediment deposition, and geomorphic structures associated with deltas.

Figure IV-3-5. Flood shoal, East Pass, Florida (23 Sep 1987). North is to the top. The Gulf of Mexico is about 1 km south of the Hwy 98 bridge at the bottom of the image. Dunes in the main channel are flood-oriented (toward the top of the image), while bed forms in the marginal channels are ebb-oriented (Photograph courtesy of USAED, Mobile)

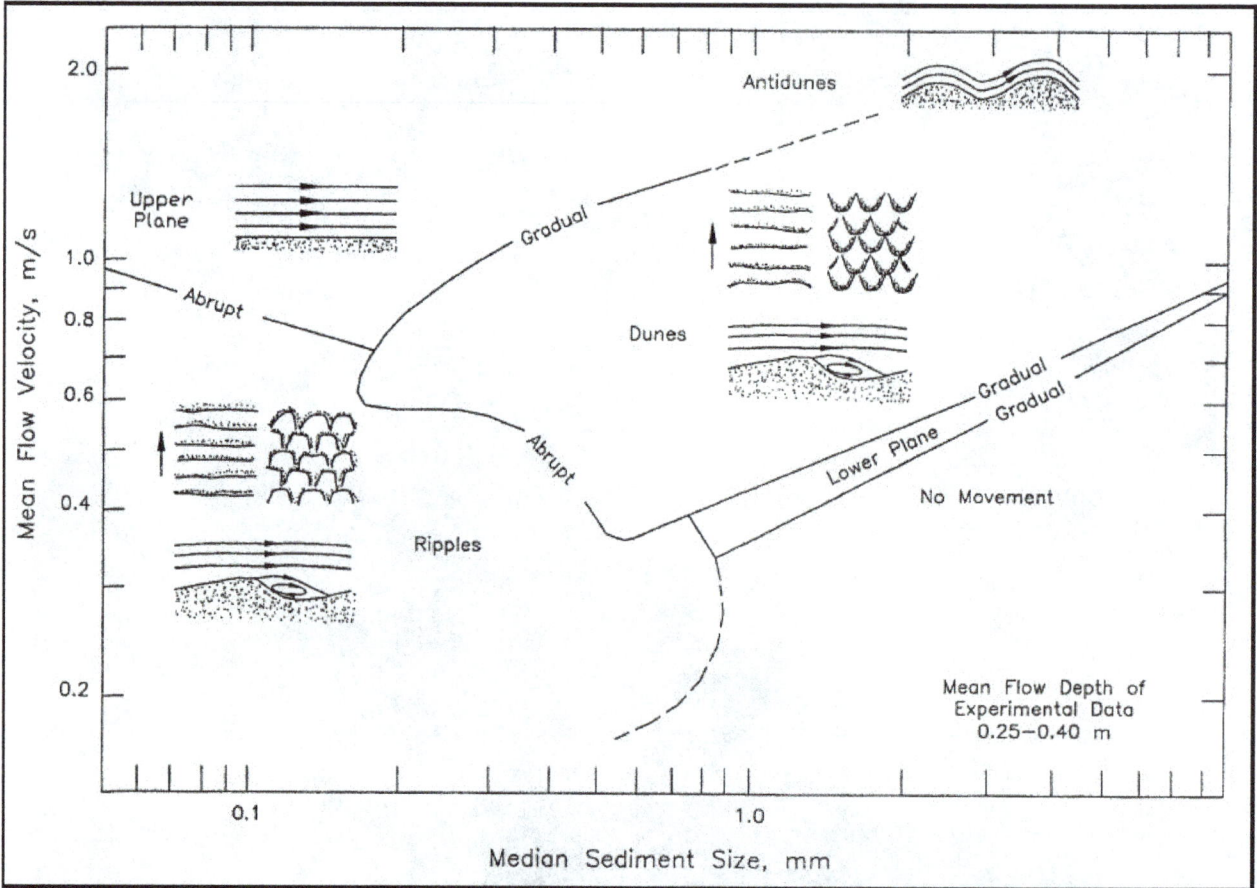

**Figure IV-3-6. Plot of mean flow velocity against mean grain size, based on laboratory studies, showing stability phases of subaqueous bed forms (modified from Ashley (1990)). Original data from various sources, standardized to 10° C water temperature (original data points not shown)**

*b. General delta classification.* Coleman and Wright (1975) identified six broad classes of deltas using an energy criteria. These models have been plotted on Figure IV-3-7 according to the relative importance of river, wave, and tide processes. However, Wright (1985) acknowledged that because each delta has unique and distinct features, no classification scheme can adequately encompass the wide variety of environments and structures found at deltas around the world.

*c. Delta-forming processes.*

(1) Force balance. Every delta is the result of a balance of forces that interact in the vicinity of the river mouth. A river carries sediment to the coast and deposits it beyond the mouth. Tidal currents and waves rework the newly deposited sediments, affecting the shape and form of the resulting structure. The long-term evolution of a delta plain becomes a function of the rate of riverine sediment input and the rate and pattern of sediment reworking, transport, and deposition by marine processes after the initial deposition. On a large scale, gross deltaic shape is also influenced by receiving basin geometry, regional tectonic stability, rates of subsidence caused by compaction of newly deposited sediment, and rate of sea level rise.

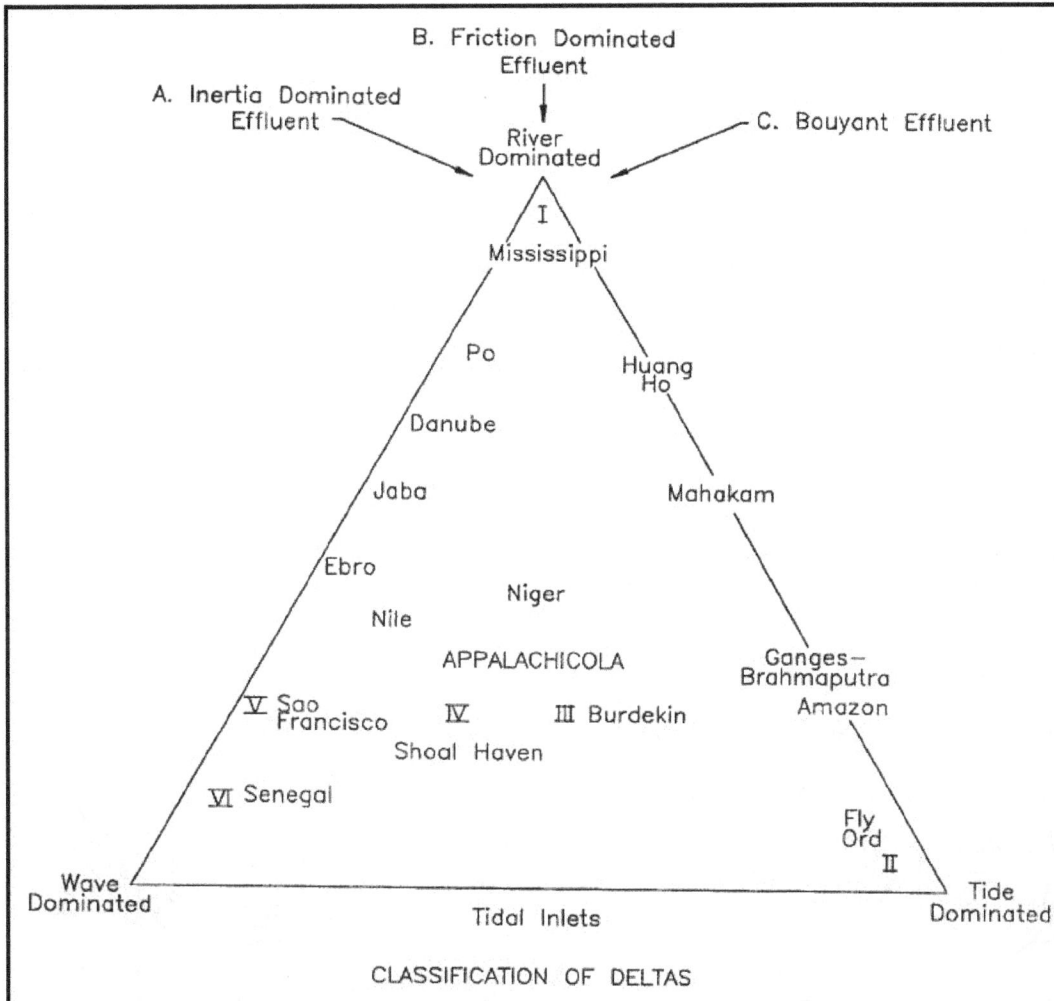

Figure IV-3-7. Classification of deltas

(2) River-dominant deltas.

(a) River-dominant deltas are found where rivers carry so much sediment to the coast that the deposition rate overwhelms the rate of reworking and removal due to local marine forces. In regions where wave energy is very low, even low-sediment-load rivers can form substantial deltas.

(b) When a river is completely dominant over marine forces, the delta shape develops as a pattern of prograding, branching distributary channels (resembling fingers branching from a hand). Interdistributary features include open bays and marshes. A generalized isopach map for this type of delta (Type I in Coleman and Wright's (1975) classification) is shown in Figure IV-3-8. A prime example is the Mississippi River, which not only transports an enormous amount of sediment, but also empties into the low wave-energy, low tide-range Gulf of Mexico (Figure IV-3-9). The Fraser River, which flows into the sheltered Strait of Georgia, is another example (Figure IV-3-10). The Mississippi is discussed in more detail later.

I. Low Wave Energy; Low Littoral
   Drift; High Suspended Load

II. Low Wave Energy; Low Littoral
    Drift; High Tide

Shoreline Trend

III. Intermediate Wave Energy;
     High Tide; Low littoral Drift

IV. Intermediate Wave Energy;
    Low Tide

V. High Wave Energy; Low Littoral
   Drift; Steep Offshore Slope

VI. High Wave Energy; High Littoral
    Drift; Steep Offshore Slope

0        16
      km

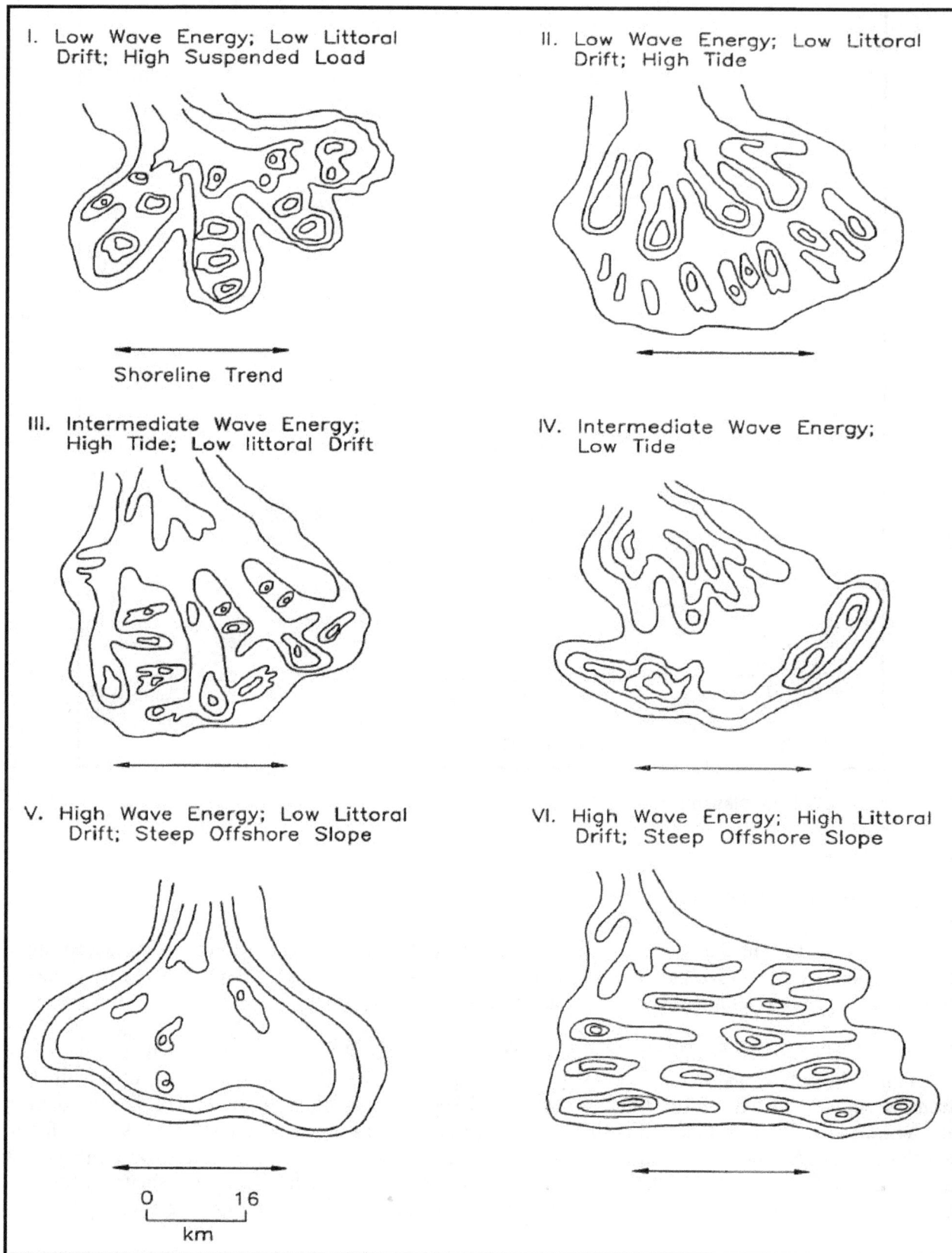

Figure IV-3-8.  Isopach map for river-dominant deltas

Coastal Morphodynamics

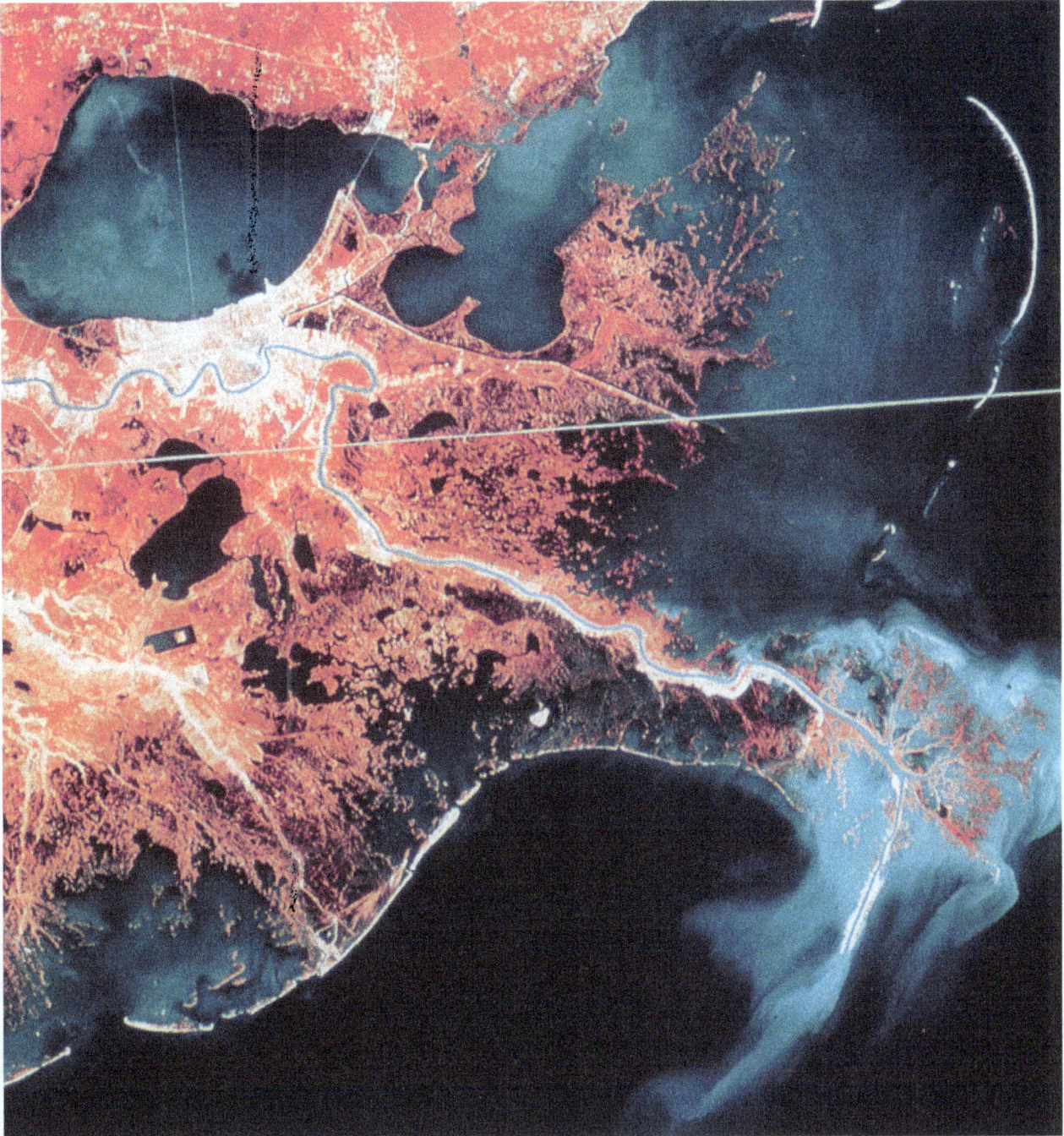

Figure IV-3-9. Mississippi River Delta. The river, which flows into the Gulf of Mexico, has several outlets that are dredged by the Corps of Engineers almost continuously. The city of New Orleans is at the upper left of the image immediately south of Lake Pontchartrain. Landsat 2 image (date unknown)

(3) Wave-dominant deltas.

(a) At wave-dominant deltas, waves sort and redistribute sediments delivered to the coast by rivers and remold them into shoreline features such as beaches, barriers, and spits. The morphology of the resulting delta reflects the balance between sediment supply and the rate of wave reworking and redistribution. Wright and Coleman (1972, 1973) found that deltas in regions of the highest nearshore wave energy flux developed

**Figure IV-3-10. Fraser River Delta, British Columbia. The river flows into the sheltered Strait of Georgia past the city of Vancouver. The strait has high tidal currents, as shown by the plume of turbid water flowing to the southeast. This delta is also Class I, riverine-dominated (NASA ERTS, 12 Aug 1973)**

the straightest shorelines and best-developed interdistributary beaches and beach-ridge complexes (Figure IV-2-3).

(b) Of 16 deltas compared by Wright and Coleman (1972, 1973), the Mississippi was the most river-dominated while the Senegal in west Africa was the other extreme, the most wave-dominated. A model of the Senegal (Type VI in Figure IV-3-8) shows that abundant beach ridges are parallel to the prevailing shoreline trend and that the shore is relatively straight as a result of high wave energy and a strong unidirectional littoral drift.

(c) An intermediate delta form is represented by the delta of the Rio São Francisco del Norte in Brazil (Type V in Figure IV-3-8). Distributary-mouth-bar deposits are restricted to the immediate vicinity of the river mouth and are quickly remolded by waves. Persistent wave energy redistributes the riverine sediment to form extensive sand sheets. The exposed delta plain consists primarily of beach ridges and aeolian dunes.

(4) Tide-dominant deltas.

Three important processes characterize tide-dominated deltas:

(a) At the river mouths, mixing obliterates vertical density stratification, eliminating the effects of buoyancy.

(b) For part of the year, tidal currents may be responsible for a greater fraction of the sediment-transporting energy than the river. As a result, sediment transport in and near the river mouth is bidirectional over a tidal cycle.

(c) The location of the land-sea interface and the zone of marine-riverine interactions is greatly extended both vertically and horizontally. Examples of deltas that are strongly influenced by tides include the Ord (Australia), Shatt-al-Arab (Iraq), Amazon (Brazil), Ganges-Brahmaputra (Bangladesh), and the Yangtze (China). Characteristic features of river mouths in macrotidal environments are bell-shaped, sand-filled channels and linear tidal sand ridges. The crests of the ridges, which have relief of 10-20 m, may be exposed at low tide. The ridges replace the distributary-mouth bars found at other deltas and become the dominant sediment-accumulation form. As the delta progrades over time, the ridges grow until they are permanently exposed, forming large, straight tidal channels (Type II in Figure IV-3-8). An example of a macrotidal delta is the Ord of Western Australia and the Essex River Delta in Massachusetts (Figure IV-3-11).

(5) Intermediate forms.

(a) As stated above, the morphology of most deltas is a result of a combination of riverine, tidal, and wave forces. One example of an intermediate form is the Burdekin Delta of Australia (Type II in Figure IV-3-8). High waves redistribute sands parallel to the coastline trend and remold them into beach ridges and barriers. Within the river mouths, tidal currents produce sand-filled river channels and tidal creeks. This type of delta displays a broad range of characteristics, depending upon the relative strength of waves versus tides. In addition, features may vary seasonally if runoff and wave climate change. Other examples include the Irrawaddy (Burma), Mekong (Vietnam), and Red (Vietnam) Deltas (Wright 1985).

(b) The fourth model of delta geometry is characterized by offshore bay-mouth barriers that shelter lagoons, bays, or estuaries into which low-energy deltas prograde (Type IV, Figure IV-3-8). Examples include the Appalachicola (Florida Panhandle), Sagavanirktok (Alaska), and Shoalhaven (southeastern Australia) Deltas (Wright 1985). In contrast to the river-dominant models, the major accumulation of prodelta mud occurs landward of the main sand body (the barrier), and at the same elevation, within the protected bay. Although suspended fines reach the open sea, wave action prevents mud accumulation as a distinct unit over the open shelf.

*d. Deltaic components and sediments.*

(1) Generally, all deltas consist of four physiographic zones: an alluvial valley, upper deltaic plain, lower deltaic plain, and subaqueous deltaic plain (Figure IV-3-12). The deposition that occurs adjacent to and between the distributory channels accounts for most of the subaerial delta. In the case of the Mississippi

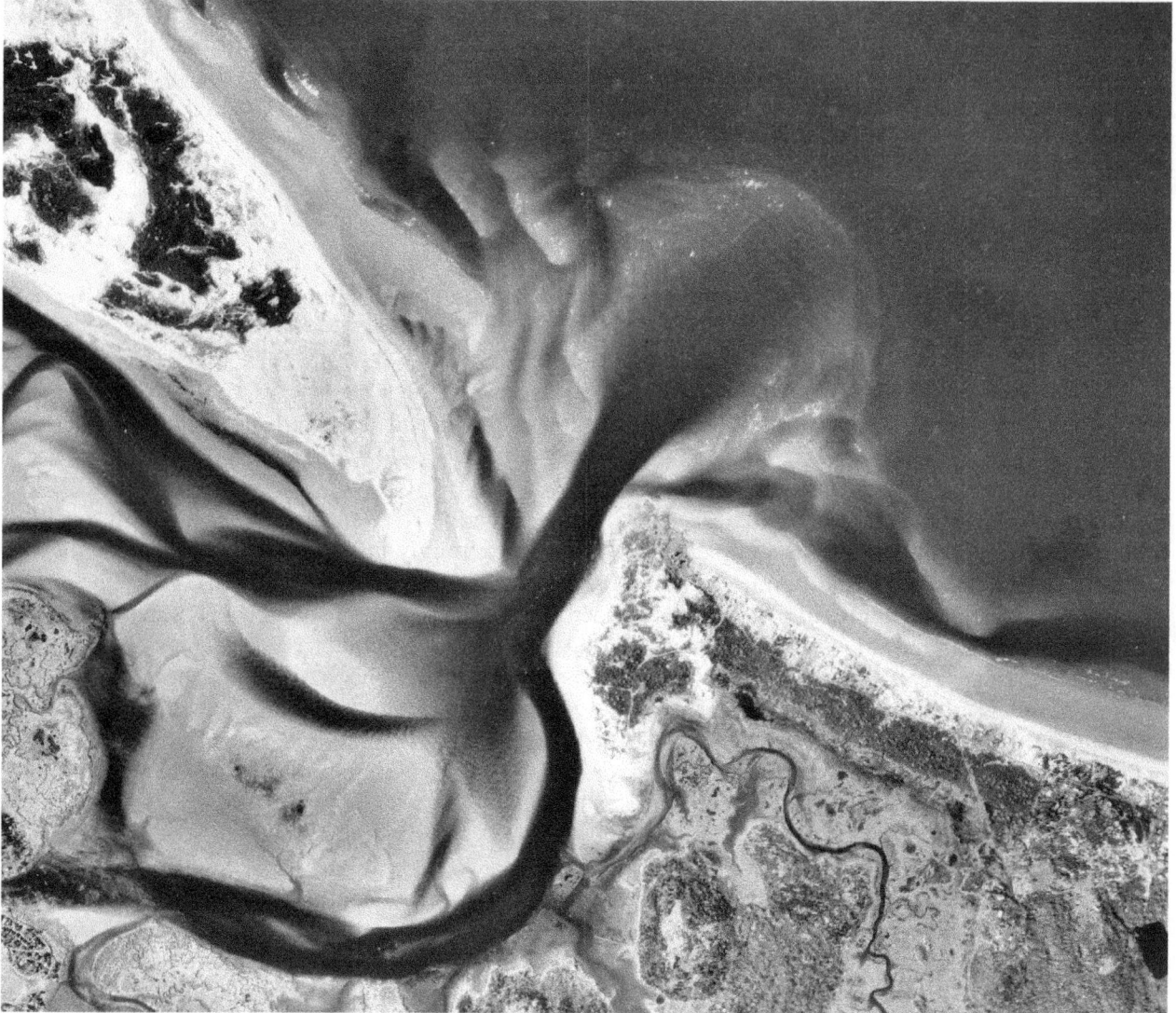

**Figure IV-3-11. Mouth of the Essex River, Massachusetts (23 April 1978). This delta is Class II, high tide and low wave energy, or possibly Class III, intermediate wave and high tide. This river mouth is anchored on the south by rock outcrops. Formerly, there may have been more open water in the back bays, and the morphology would have resembled an inlet in a barrier spit**

delta, significant sand accumulates in the interdistributory region when breaks in the levees occur, allowing river water to temporarily escape from the main channel. These accumulations are called *crevasse splays*.

(2) The subaqueous plain is the foundation over which the modern delta progrades (as long as the river occupies the existing course and continues to supply sufficient sediment). The subaqueous plain is characterized by a seaward-fining of sediments, with sand being deposited near the river mouths and clays settling further offshore. The seawardmost unit of the plain is the prodelta. It overlies the sediments of the inner continental shelf and consists of a blanket of clays deposited from suspension. The prodelta of the Mississippi ranges from 20 to 50 m thick and extends seaward to water depths of 70 m. The Mississippi's prodelta contains pods of distributory mouth bar sands and their associated cross bedding, flow structures, and shallow-water fauna. These pods may be slump blocks carried down to the prodelta by submarine landslides (Prior and Coleman 1979). Slumping and mudflow are mechanisms that transport large masses of sediment down to the edge of the continental slope and possibly beyond. These mass movements are a

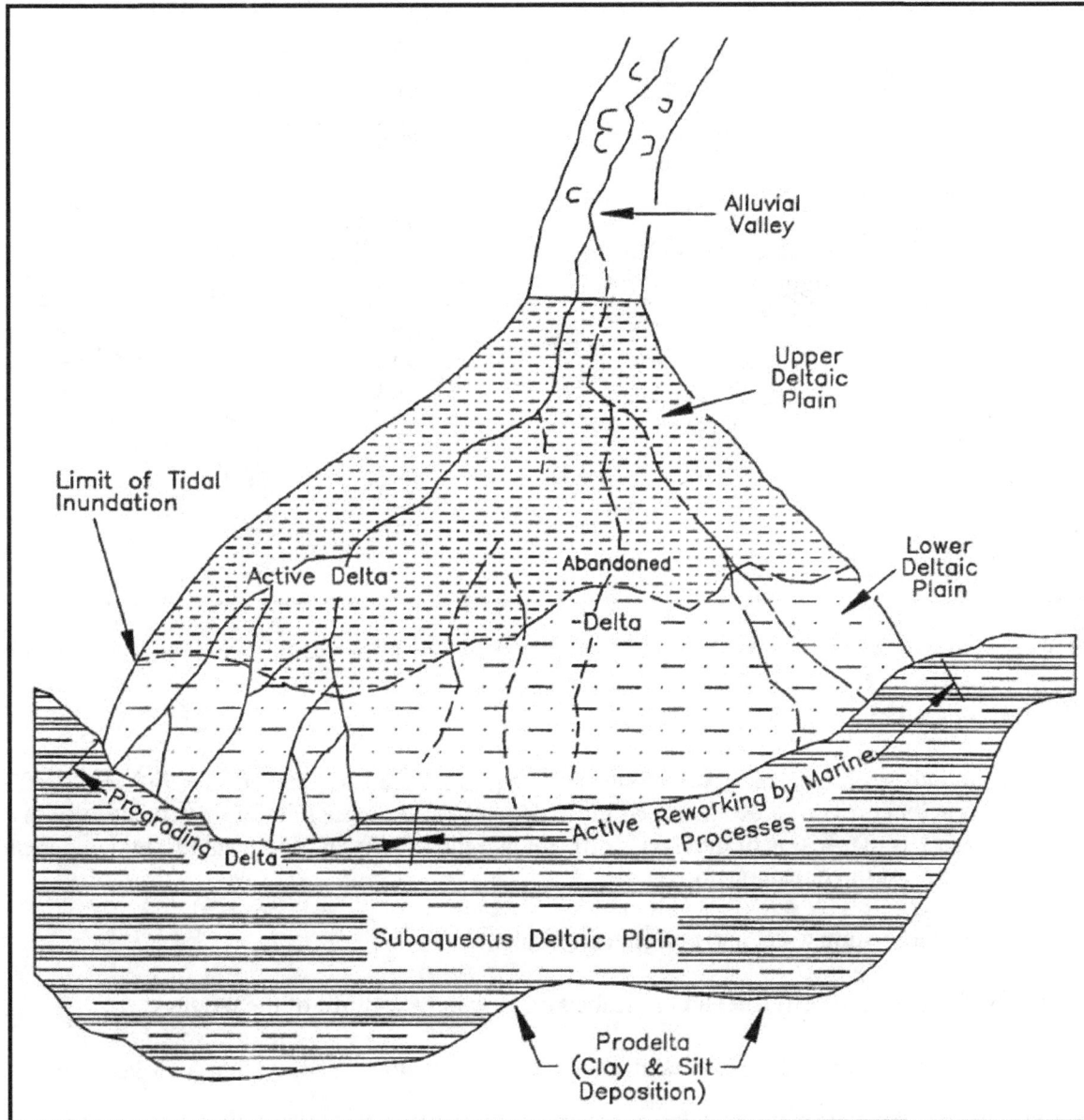

Figure IV-3-12. Basic physiographic units common to all deltas (from Wright (1985))

serious hazard to oil drilling and production platforms. Mud diapirs, growth faults, mud/gas vents, pressure ridges, and mudflow gullies are other evidence of sediment instability on the Mississippi Delta (Figure IV-3-13). Additional details of this interesting subject are covered in Coleman (1988), Coleman and Garrison (1977), Henkel (1970), and Prior and Coleman (1980).

(3) Above the delta front, there is a tremendous variability of sediment types. A combination of shallow marine processes, riverine influence, and brackish-water faunal activity causes the interdistributory bays to display an extreme range of lithologic and textural types. On deltas in high tide regions, the interdistributory bay deposits are replaced by tidal and intertidal flats. West of the Mississippi Delta is an extensive chenier plain. Cheniers are long sets of sand beach ridges, located on mudflats.

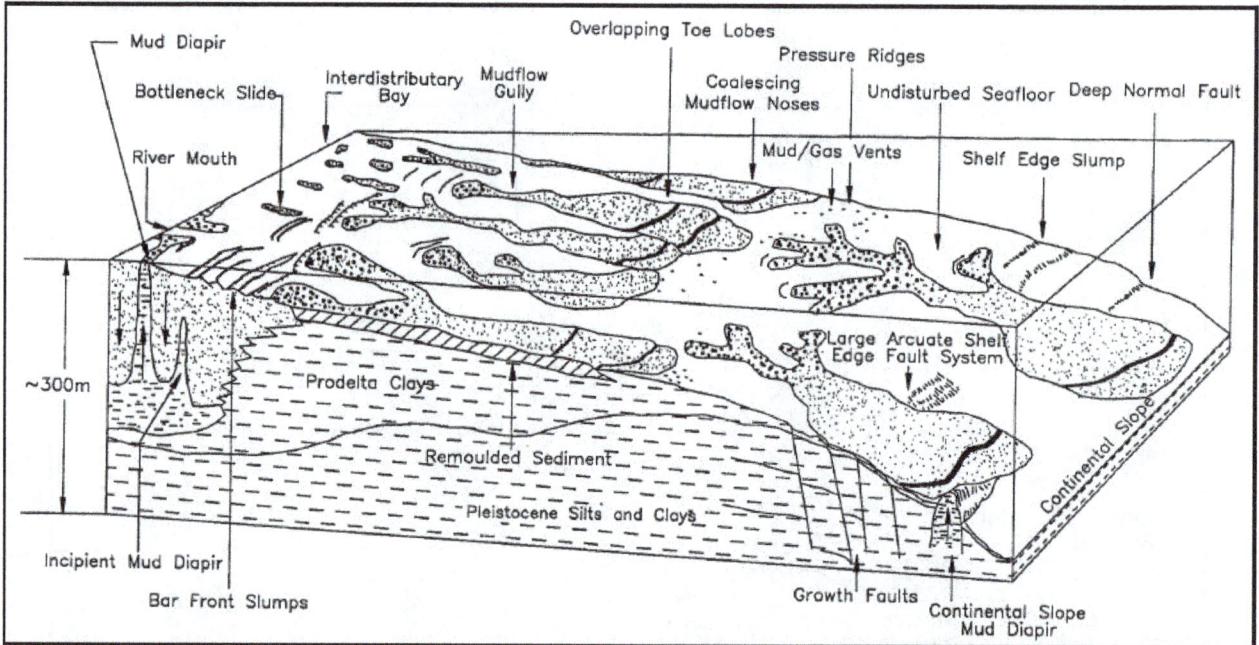

**Figure IV-3-13. Structures and types of sediment instabilities on the Mississippi Delta (from Coleman (1988))**

*e. River mouth flow and sediment deposition.*

(1) River mouth geometry and river mouth bars are influenced by, and in turn influence, effluent dynamics. This subject needs to be examined in detail because the principles are pertinent to both river mouths and tidal inlets. Diffusion of the river's effluent and the subsequent sediment dispersion depend on the relative strengths of three main factors:

(a) Inertia of the issuing water and associated turbulent diffusion.

(b) Friction between the effluent and the seabed immediately seaward of the mouth.

(c) Buoyancy resulting from density contrasts between river flow and ambient sea or lake water.

Based on these forces, three sub-classes of deltaic deposition have been identified for river-dominated deltas (Figure IV-3-7). Two of these are well illustrated by depositional features found on the Mississippi Delta.

(2) Depositional model type A - inertia-dominated effluent.

(a) When outflow velocities are high, depths immediately seaward of the mouth tend to be large, density contrasts between the outflow and ambient water are low, and inertial forces dominate. As a result, the effluent spreads and diffuses as a turbulent jet (Figure IV-3-14a). As the jet expands, its momentum decreases, causing a reduction of its sediment carrying capacity. Sediments are deposited in a radial pattern, with the coarser bed load dropping just beyond the point where the effluent expansion is initiated. The result is basinward-dipping foreset beds.

(b) This ideal model is probably unstable under most natural conditions. As the river continues to discharge sediment into the receiving basin, shoaling eventually occurs in the region immediately beyond the mouth (Figure IV-3-14b). For this reason, under typical natural conditions, basin depths in the zone of the

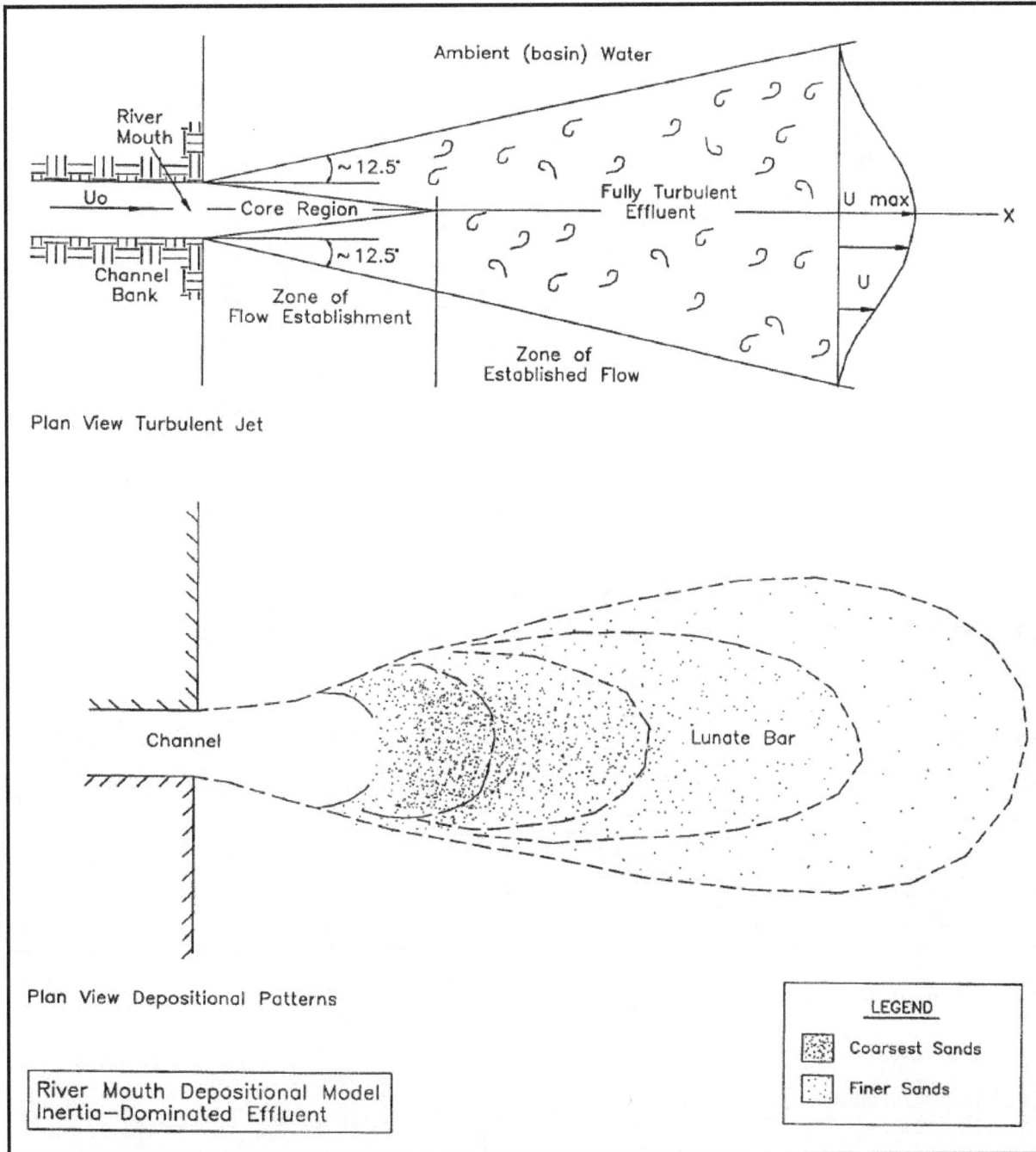

Figure IV-3-14. Plan view of depositional Model A, inertia-dominated effluent (adapted from Wright (1985)) (Continued)

jet's diffusion are unlikely to be deeper than the outlet depth. Effluent expansion and diffusion become restricted horizontally as a plane jet. More important, friction becomes a major factor in causing rapid deceleration of the jet. Model 'A' eventually changes into friction-dominated Model 'B'.

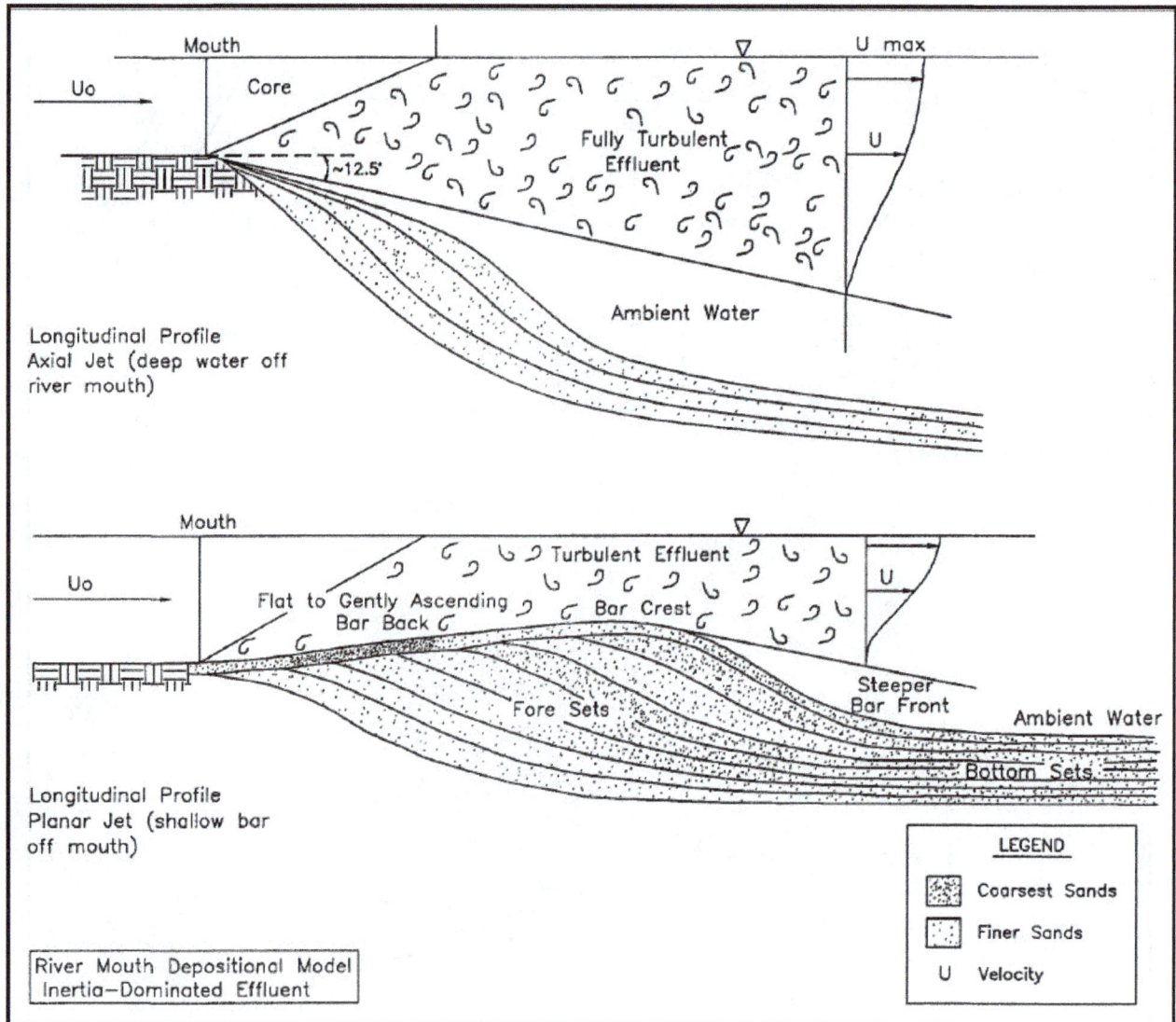

**Figure IV-3-14. (Concluded)**

(3) Depositional model type B - friction-dominated effluent.

(a) When homopycnal,[1] friction-dominated outflow issues over a shallow basin, a distinct pattern of bars and subaqueous levees is formed (Figure IV-3-15). Initially, the rapid expansion of the jet produces a broad, arcuate radial bar. As deposition continues, natural subaqueous levees form beneath the lateral boundaries of the expanding jet where the velocity decreases most rapidly. These levees constrict the jet from expanding further. As the central portion of the bar grows upward, channels form along the lines of greatest turbulence, which tend to follow the subaqueous levees. The result is the formation of a bifurcating channel that has a triangular middle-ground shoal separating the diverging channel arms. The flow tends to be concentrated into the divergent channels and to be tranquil over the middle ground under normal conditions.

---

[1] River water and ambient water have the same density (for example, a stream entering a freshwater lake).

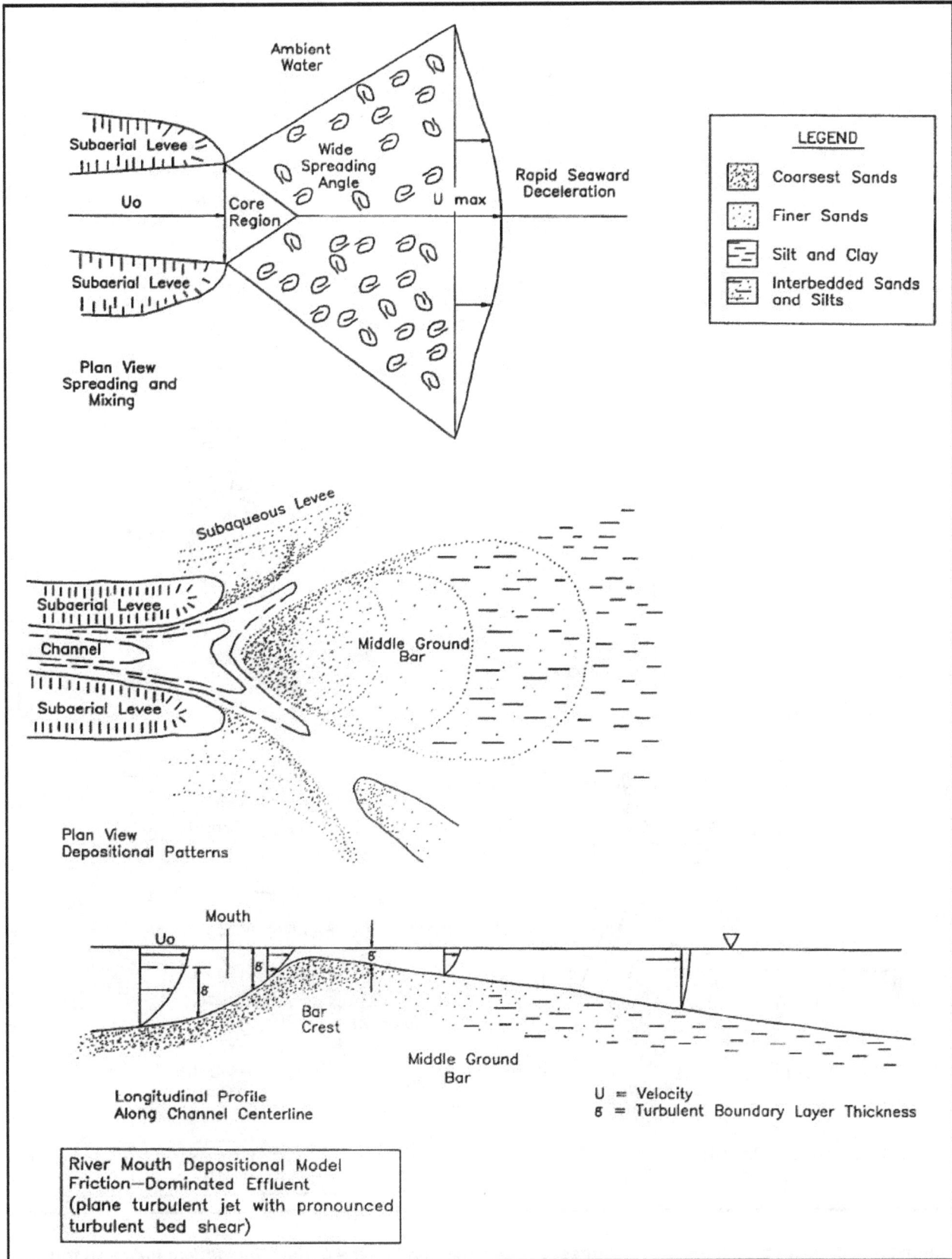

Figure IV-3-15. Depositional model type B, friction-dominated effluent (adapted from Wright (1985))

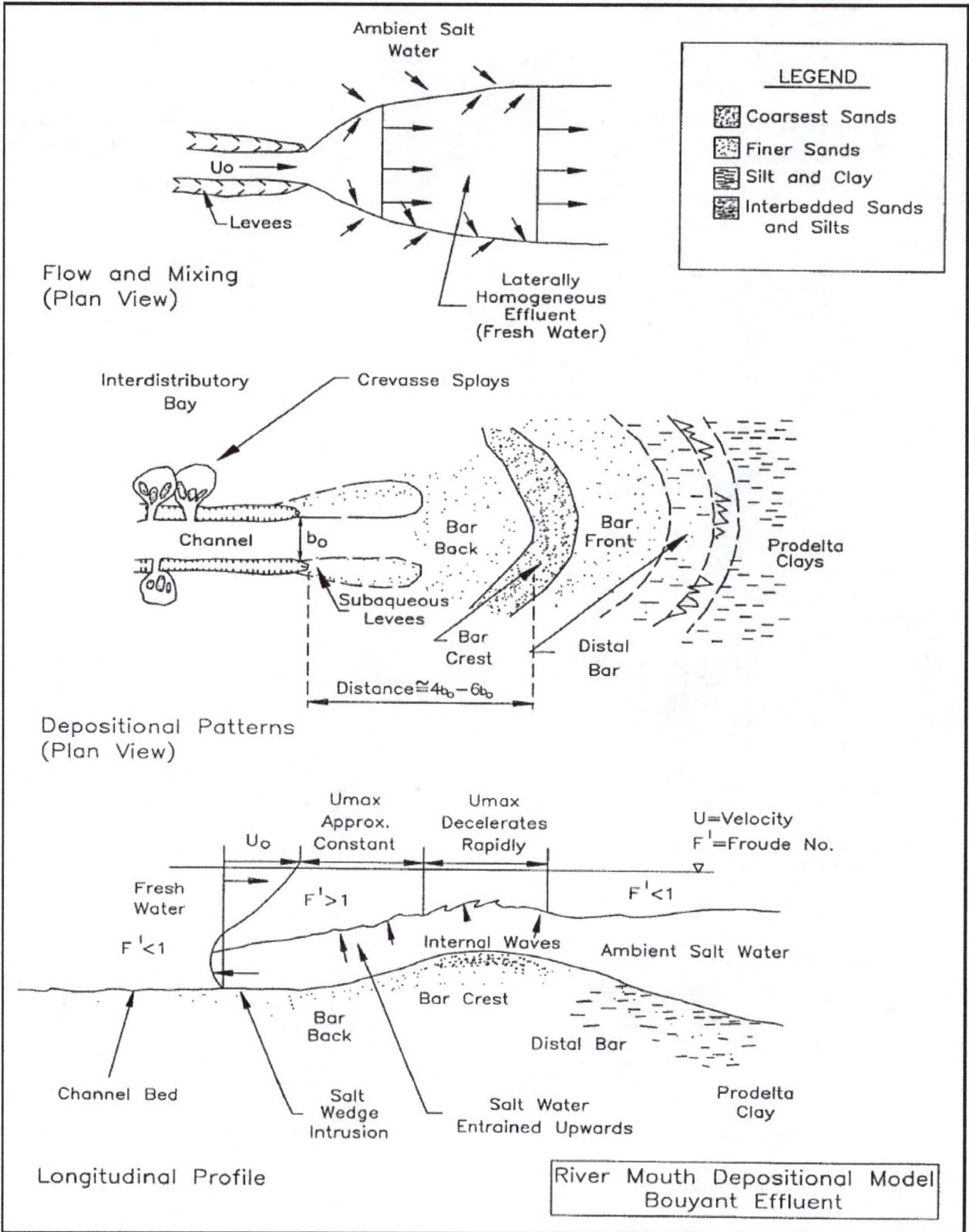

Figure IV-3-16. River mouth bar crest features, depositional model type C, buoyant effluent (adapted from Wright (1985)) (Continued)

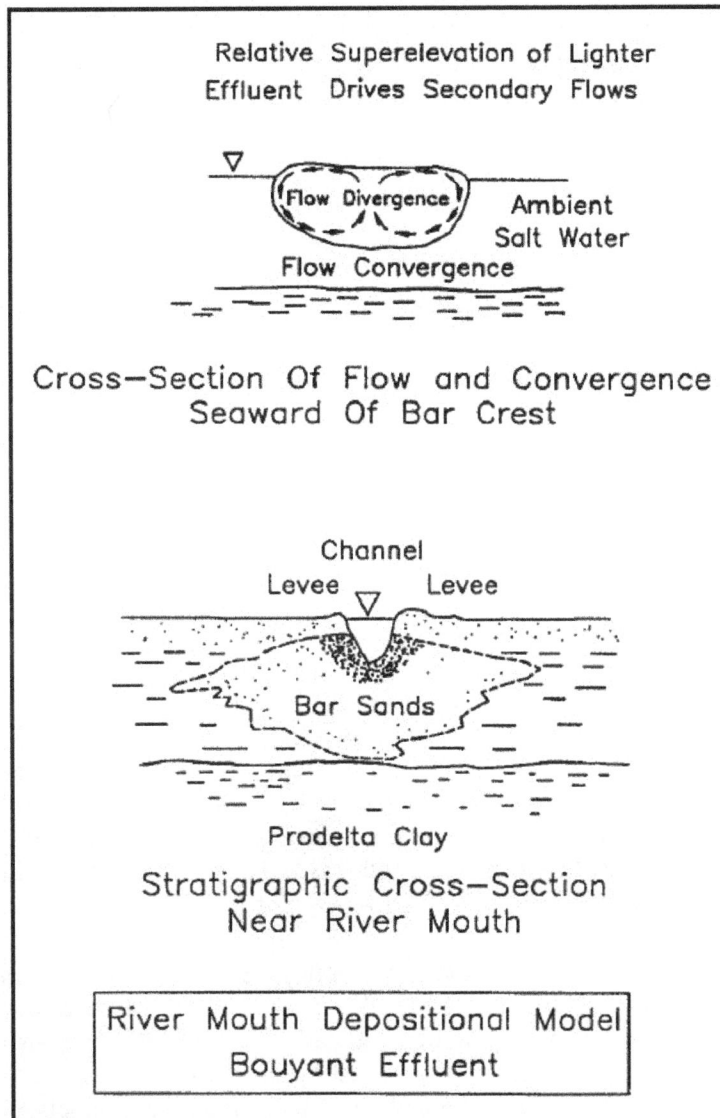

**Figure IV-3-16.  (Concluded)**

(b) This type of bar pattern is most common where nonstratified outflow enters a shallow basin. Examples of this pattern (known as *crevasse splays* or *overbank splays*) are found at crevasses along the Mississippi River levees.  These secondary channels run perpendicular to the main Mississippi channels and allow river water to debouch into the broad, shallow interdistributory bays.  This process forms the major sub-aerial land (marsh) of the lower Mississippi delta (Coleman 1988).

(4)  Depositional model type C - buoyant effluent.

(a)  Stratification often occurs when fresh water flows out into a saline basin. When the salt wedge is well developed, the effluent is effectively isolated from the effects of bottom friction. Buoyancy suppresses mixing and the effluent spreads over a broad area, thinning progressively away from the river mouth (Figure IV-3-16).  Deceleration of the velocity of the effluent is caused by the upward entrainment of seawater across the density interface.

(b) The density interface between the freshwater plume and the salt wedge is often irregular due to internal waves (Figure IV-3-16a). The extent that the effluent behaves as a turbulent or buoyant jet depends largely on the Froude number $F'$:

$$F' = \frac{U^2}{\gamma g h'}$$

(IV-3-1)

where

$U$ = mean outflow velocity of upper layer (in case of stratified flow)

$g$ = acceleration of gravity

$h'$ = depth of density interface

$$\gamma = 1 - (\rho_f/\rho_s)$$

(IV-3-2)

where

$\rho_f$ = density of fresh water

$\rho_s$ = density of salt water

As $F'$ increases, inertial forces dominate, accompanied by an increase in turbulent diffusion. As $F'$ decreases, turbulence decreases and buoyancy becomes more important. Turbulence is suppressed when $F'$ is less than 1.0 and generally increases as $F'$ increases beyond 1.0 (Wright 1985).

(c) The typical depositional patterns associated with buoyant effluent are well represented by the mouths of the Mississippi River (Wright and Coleman 1975). Weak convergence near the base of the effluent inhibits lateral dispersal of sand, resulting in narrow bar deposits that prograde seaward as laterally restricted "bar-finger sands" (Figure IV-3-16b). The same processes presumably prevent the subaqueous levees from diverging, causing narrow, deep distributory channels. Because the active channels scour into the underlying distributory-mouth bar sands as they prograde, accumulations of channel sands are usually limited. Once the channels are abandoned, they tend to fill with silts and clays. It is believed that the back bar and bar crest grow mostly from bed-load transport during flood stages. The subaqueous levees, however, appear to grow year-round because of the near-bottom convergence that takes place during low and normal river stages.

*f. Mississippi Delta - Holocene history, dynamic changes.*

(1) General. The Mississippi River, which drains a basin covering 41 percent of the continental United States (3,344,000 sq km), has deposited an enormous mound of unconsolidated sediment in the Gulf of Mexico. The river has been active since at least late Jurassic times and has dominated deposition in the northern Gulf of Mexico. Many studies have been conducted on the Mississippi Delta, leading to much of our knowledge of deltaic sedimentation and structure. The ongoing research is a consequence of the river's critical importance to commerce and extensive petroleum exploration and production in the northern Gulf of Mexico during the last 50 years.

(2) Deposition time scales. The Mississippi Delta consists of overlapping deltaic lobes. Each lobe covers 30,000 sq km and has an average thickness of about 35 m. The lobes represent the major sites of the river's deposition. The process of switching from an existing lobe to a new outlet takes about 1500 years

(Coleman 1988). Within a single lobe, deposition in the bays occurs from overbank flows, crevasse splays, and biological production. The bay fills, which cover areas of 250 sq km and have a thickness of only 15 m, accumulate in only about 150 years. Overbank splays, which cover areas of 2 sq km and are 3 m thick, occur during major floods when the natural levees are breached. The mouths of the Mississippi River have prograded seawards at remarkable rates. The distributory channels can form sand bodies that are 17 km long, 8 km wide, and over 80 m thick in only 200 years (Coleman 1988).

(3) Holocene history. During the last low sea level stand, 18,000 years ago, the Mississippi River entrenched its valley, numerous channels were scoured across the continental shelf, and deltas were formed near the shelf edge (Suter and Berryhill 1985). As sea level rose, the site of deposition moved upstream to the alluvial valley. By about 9,000 years before present, the river began to form its modern delta. In more recent times, the shifting deltas of the Mississippi have built a delta plain covering a total area of 28,500 sq km. The delta switching, which has occurred at high frequency, combined with a rapidly subsiding basin, has resulted in vertically stacked cyclic sequences. Because of rapid deposition and switching, in a short time the stacked cyclic deltaic sequences have attained thicknesses of thousands of meters and covered an area greater than 150,000 sq km (Coleman 1988). Figure IV-3-17 outlines six major lobes during the last 7,500 years.

(4) Modern delta. The modern delta, the Balize or Birdfoot, began to prograde about 800 to 1,000 years ago. Its rate of progradation has diminished recently and the river is presently seeking a new site of deposition. Within the last 100 years, a new distributory, the Atchafalaya, has begun to divert an increasing amount of the river's flow. Without river control structures, the new channel would by now have captured all of the Mississippi River's flow, leading to rapid erosion of the Balize Delta. (It is likely that there would be a commensurate deterioration of the economy of New Orleans if it lost its river.) Even with river control projects, the Atchafalaya is actively building a delta in Atchafalaya Bay (lobe 6 in Figure IV-3-17).

*g. Sea level rise and deltas.*

(1) Deltas experience rapid local relative sea level rise because of the natural compaction of deltaic sediments from dewatering and consolidation. Deltas are extremely vulnerable to storms because the subaerial surfaces are flat and only slightly above the local mean sea level. Only a slight rise in sea level can extend the zone subject to storm surges and waves further inland. As stated earlier, delta evolution is a balance between the accumulation of fluvially supplied sediment and the reworking, erosion, and transport of deltaic sediment by marine processes (Wright 1985). Even a river like the Mississippi, which has a high sediment load and drains into a low wave-energy basin, is prograding only in the vicinity of the present distributory channels, the area defined as the active delta (Figures IV-3-9 and IV-3-12).

(2) Deltas are highly fertile agriculturally because of the steady supply of nutrient-laden soil. As a result, some of the world's greatest population densities - over 200 inhabitants per square kilometer - are found on deltas (*The Times Atlas of the World* 1980):

(a) Nile Delta, Egypt.

(b) Chang Jiang (Yangtze), China.

(c) Mekong, Vietnam.

(d) Ganga (Ganges), Bangladesh.

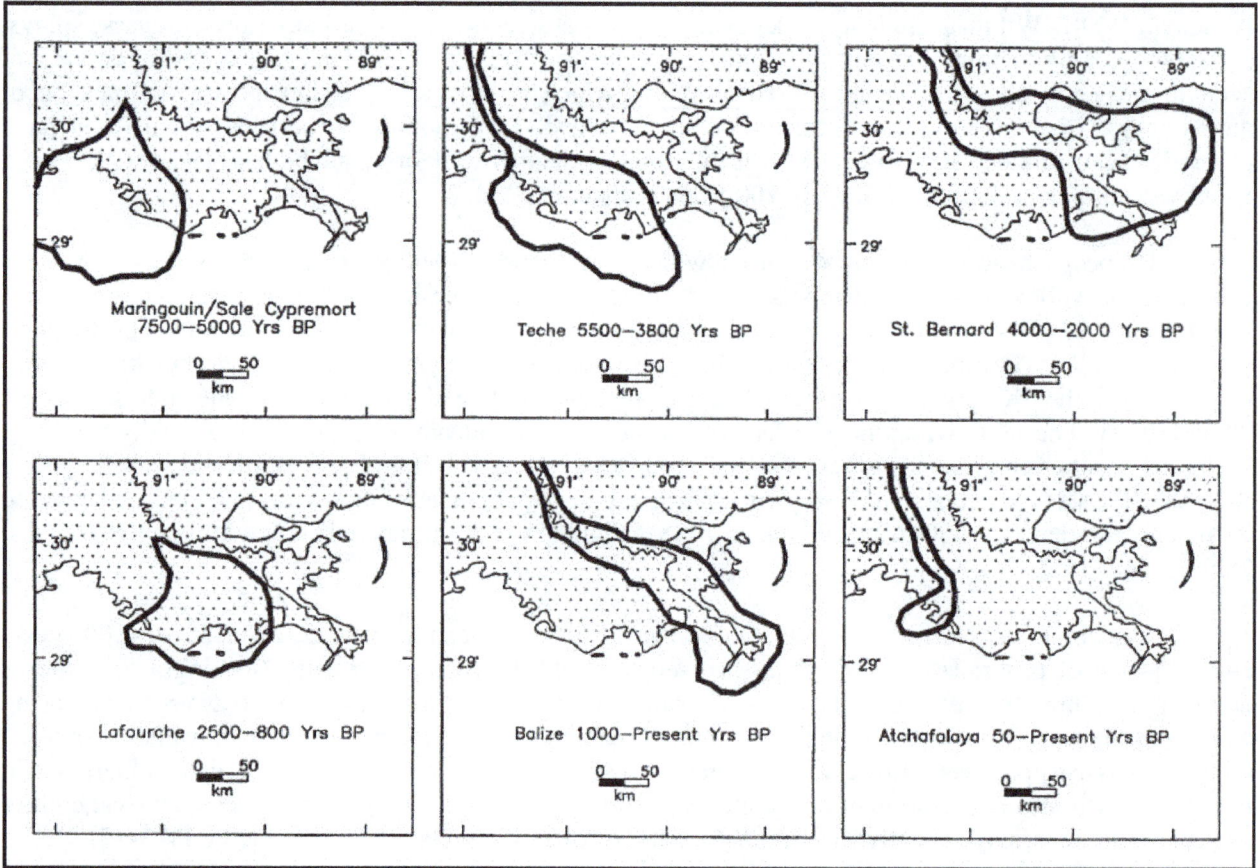

**Figure IV-3-17. Shifting sites of deltaic sedimentation of the Mississippi River (from Coleman (1988))**

These populations are very vulnerable to delta land loss caused by rising sea level and by changes in sediment supply due to natural movements of river channels or by upland man-made water control projects.

(3) Inhabitants of deltas are also in danger of short-term changes in sea level caused by storms. Tropical storms can be devastating: the Bay of Bengal cyclone of November 12, 1970, drowned over 200,000 persons in what is now Bangladesh (Carter 1988). Hopefully, public education, improving communications, better roads, and early warning systems will be able to prevent another disaster of this magnitude. Coastal management in western Europe, the United States, and Japan is oriented towards the orderly evacuation of populations in low-lying areas and has greatly reduced storm-related deaths. In contrast to the Bay of Bengal disaster, Hurricane Camille (August 17-20, 1969), caused only 236 deaths in Louisiana, Mississippi, Alabama, and Florida.

## IV-3-4. Coastal Inlets

*a. Introduction.*

(1) Coastal inlets play an important role in nearshore processes around the world. *Inlets* are the openings in coastal barriers through which water, sediments, nutrients, planktonic organisms, and pollutants are exchanged between the open sea and the protected embayments behind the barriers. In the United States, the classical image of an inlet is of an opening in one of the Atlantic or Gulf of Mexico barrier islands, but inlets are certainly not restricted to barrier environments or to shores with tides. On the West Coast and in the Great

Lakes, many river mouths are considered to be inlets, while in the Gulf of Mexico, the wide openings between the barriers, locally known as passes, are also inlets. Inlets can be cut through unconsolidated shoals or emergent barriers as well as through clay, rock, or organic reefs (Price 1968). There is no simple, restrictive definition of inlet; based on the geologic literature and on regional terminology, almost any opening in the coast, ranging from a few meters to several kilometers wide, can be called an inlet. Inlets are important economically to many coastal nations because harbors are often located in the back bays, requiring that the inlets be maintained for commercial navigation. At many inlets, the greatest maintenance cost is incurred by repetitive dredging of the navigation channel. Because inlets are hydrodynamically very complex, predictions of shoaling and sedimentation have often been unsatisfactory. A better understanding of inlet sedimentation patterns and their relationship to tidal and other hydraulic processes can hopefully contribute to better management and engineering design.

(2) Tidal inlets are analogous to river mouths in that sediment transport and deposition patterns in both cases reflect the interaction of outflow inertia and associated turbulence, bottom friction, buoyancy caused by density stratification, and the energy regime of the receiving body of water (Wright and Sonu 1975). However, two major differences usually distinguish lagoonal inlets from river mouths, sometimes known as fluvial or riverine inlets (Oertel 1982).

(a) Lagoonal tidal inlets experience diurnal or semidiurnal flow reversals.

(b) Lagoonal inlets have two opposite-facing mouths, one seaward and the other lagoonward. The sedimentary structures which form at the two openings differ because of differing energy, water density, and geometric factors.

(3) The term *lagoon* refers to the coastal pond or embayment that is connected to the open sea by a tidal inlet. The *throat* of the inlet is the zone of smallest cross section, which, accordingly, has the highest flow velocities. The *gorge* is the deepest part of an inlet and may extend seaward and landward of the throat (Oertel 1988). *Shoal* and *delta* are often used interchangeably to describe the ebb-tidal sand body located at the seaward mouth of an inlet.

  b.  *Technical literature.* Pioneering research on the stability of inlets was performed by Francis Escoffier (1940, 1977). O'Brien (1931, 1976) derived general empirical relationships between tidal inlet dimension and tidal prism. Keulegan (1967) developed algorithms to relate tidal prism to inlet cross section. Bruun (1966) examined inlets and littoral drift, and Bruun and Gerritsen (1959, 1961) studied bypassing and the stability of inlets. Hubbard, Oertel, and Nummedal (1979) described the influence of waves and tidal currents on tidal inlets in the Carolinas and Georgia. Hundreds of other works are referenced in the USACE *General Investigation of Tidal Inlets* (GITI) reports (Barwis 1976), in special volumes like *Hydrodynamics and Sediment Dynamics of Tidal Inlets* (Aubrey and Weishar 1988), in textbooks on coastal environments (Carter 1988; Cronin 1975; Komar 1998), and in review papers (Boothroyd 1985; FitzGerald 1988). Older papers on engineering aspects of inlets are cited in Castañer (1971). There are also numerous foreign works on tidal inlets: Carter (1988) cites references from the British Isles; Sha (1990) from the Netherlands; Nummedal and Penland (1981) and FitzGerald, Penland, and Nummedal (1984) from the North Sea coast of Germany; and Hume and Herdendorf (1988, 1992) from New Zealand. More references are listed in Parts II-6 and V-6.

  c.  *Classification of inlets and geographic distribution.*

(1) Tidal inlets are found around the world in a broad range of sizes and shapes. Because of their diversity, it has been difficult to develop a suitable classification scheme. One approach has been to use an energy-based criteria, in which inlets are ranked according to the wave energy and tidal range of the environment in which they are located (Figure IV-1-10).

(2) Regional geological setting can be a limiting factor restricting barrier and, in turn, inlet development. Most inlets are on trailing-edge coasts with wide coastal plains and shallow continental shelves (e.g., the Atlantic and Gulf of Mexico shores of the United States). High relief, leading-edge coastlines have little room for sediment to accumulate either above or below sea level. Sediment tends to collect in pockets between headlands, few lagoons are formed, and inlets are usually restricted to river mouths. The Pacific coast of North America, in addition to being steep, is subject to high wave energy and has far fewer inlets than the Atlantic.

(3) Underlying geology may also control inlet location and stability. Price and Parker (1979) reported that certain areas along the Texas coast were always characterized by inlets, although the passes tended to migrate back and forth along a limited stretch of the shore. The positions of these permanent inlets were tectonically controlled, but the openings were maintained by tidal harmonics and hydraulics. If storm inlets across barriers were not located at the established stable pass areas, the inlets usually closed quickly. Some inlets in New England are anchored by bedrock outcrops and therefore cannot move freely (for example, the Essex River mouth, Figure IV-3-11).

  *d.   Hydrodynamic processes in inlets.* See Part II-6.

  *e.   Geomorphology of tidal inlets.* Tidal inlets are characterized by large sand bodies that are deposited and shaped by tidal currents and waves. The *ebb-tide shoal* (or delta) is a sand mass that accumulates seaward of the mouth of the inlet. It is formed by ebb tidal currents and is modified by wave action. The *flood-tide shoal* is an accumulation of sand at the landward opening of an inlet that is mainly shaped by flood currents (Figure IV-3-18). Depending on the size and depth of the bay, a flood shoal may extend into open water or may merge into a complex of meandering tributary channels, point bars, and muddy estuarine sediments.

  (1) Ebb-tidal deltas (shoals).

  (a) A simplified morphological model of a natural (unjettied) ebb-tidal delta is shown in Figure IV-3-18. The delta is formed from a combination of sand eroded from the gorge of the inlet and sand supplied by longshore currents. This model includes several components:

- A main *ebb channel*, scoured by the ebb jets.

- *Linear* bars that flank the main channel, the result of wave and tidal current interaction.

- A *terminal lobe*, located at the seaward (distal) end of the ebb channel. This is the zone where the ebb jet velocity drops, resulting in sediment deposition.

- *Swash platforms*, which are sand sheets located between the main ebb channel and the adjacent barrier islands.

- *Swash bars* that form and migrate across the swash platforms because of currents (the swash) generated by breaking waves.

- *Marginal flood channels*, which often flank both updrift and downdrift barriers.

Inlets with jetties often display these components, although the marginal flood channels are usually missing.

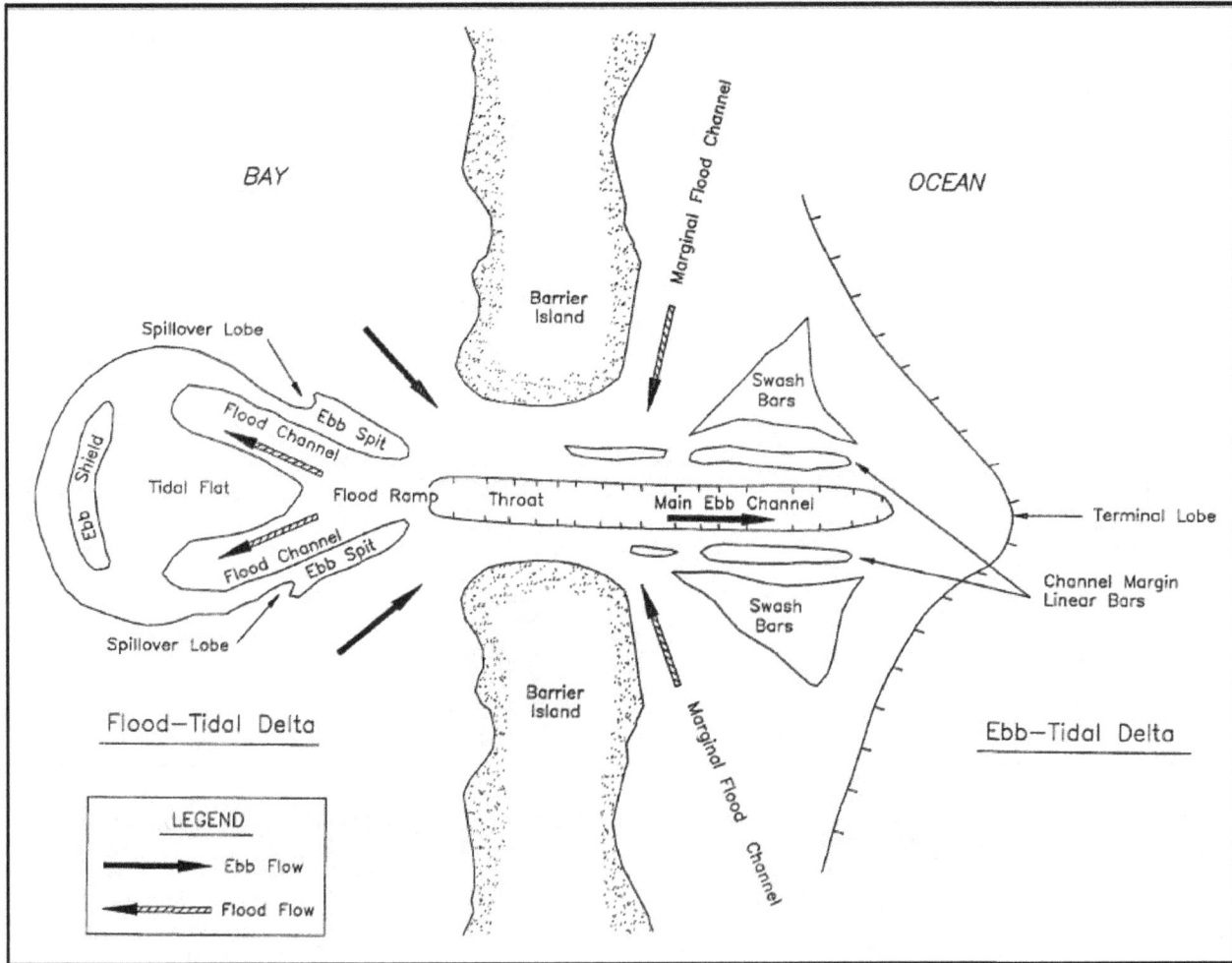

**Figure IV-3-18. Definition diagram of a tidal inlet with well-developed flood and ebb deltas (from Boothroyd (1985) and other sources)**

(b) For the Georgia coast, Oertel (1988) described simple models of ebb-delta shape and orientation which depended on the balance of currents (Figure IV-3-19). With modifications, these models could apply to most inlets. When longshore currents were approximately balanced and flood currents exceeded ebb, a squat, symmetrical delta developed (Figure IV-3-19a) (example: Panama City, Florida). If the prevailing longshore currents exceeded the other components, the delta developed a distinct northerly or southerly orientation (Figures IV-3-19b and 19c). Note that some of the Georgia ebb deltas change their orientation seasonally, trending north for part of the year and south for the rest. Finally, when inlet currents exceeded the forces of longshore currents, the delta was narrower and extended further out to sea (Figure IV-3-19d) (example: Brunswick, Georgia).

(c) Based on studies of the German and Georgia bights, Nummedal and Fischer (1978) concluded that three factors were critical in determining the geometry of the inlet entrance and the associated sand shoals:

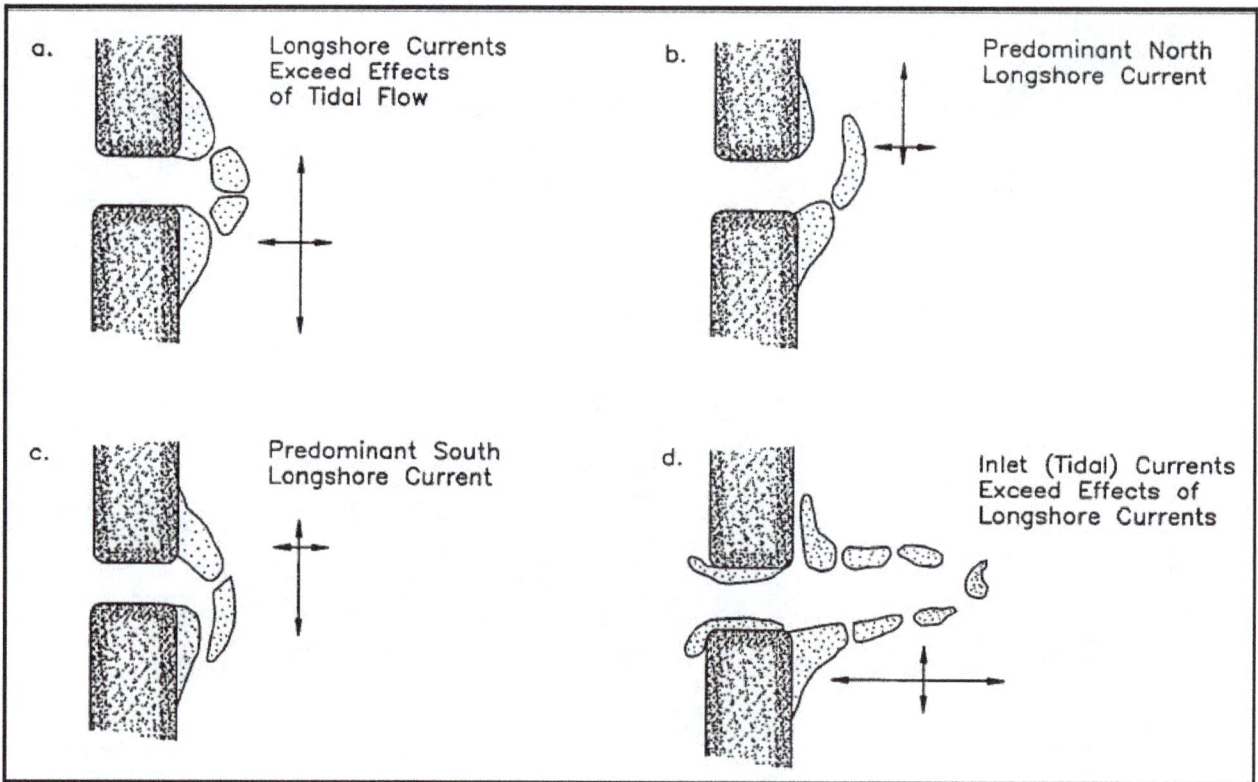

**Figure IV-3-19. Four different shapes of ebb-tidal deltas, modified by the relative effects of longshore versus tidal currents (from Oertel (1988))**

- Tide range.

- Nearshore wave energy.

- Bathymetry of the back-barrier bay.

For the German and Georgia bights, the latter factor controls velocity asymmetry through the inlet gorge, resulting in greater seaward-directed sediment transport through the inlet than landward transport. This factor has aided the development of large ebb shoals along these coasts, even though the German coast is subject to high wave energy. Back bay area and geometry are likely crucial factors that need to be incorporated in a comprehensive inlet classification scheme.

(d) Net sediment movement. At Price Inlet, South Carolina (FitzGerald and Nummedal 1983) and North Inlet, South Carolina (Nummedal and Humphries 1978), because of peak ebb currents, the resulting seaward-directed sediment transport far exceeded the sediment moved landward during flood. However, ebb velocity dominance does not necessarily mean that net sediment movement is also seaward. At Sebastian Inlet, on Florida's east coast, Stauble et al. (1988) found that net sediment movement was landward although the tidal hydraulics displayed higher ebb currents. The authors concluded that sediment carried into the inlet with the flood tide was deposited on the large, and growing, flood shoal. During ebb tide, current velocities over the flood shoal were too low to remobilize as much sediment as had been deposited on the shoal by the flood tide. The threshold for sediment transport was not reached until the flow was in the relatively narrow throat. In this case, the flood shoal had become a sink for sediment carried into the inlet. Stauble et al. (1988) hypothesized that this pattern of net landward sediment movement, despite ebb hydraulic dominance, may occur at other inlets in microtidal shores that open into large lagoons.

(e) The ebb-tidal deltas along mixed-energy coasts (e.g., East and West Friesian Islands of Germany, South Carolina, Georgia, Virginia, and Massachusetts) are huge reservoirs of sand. FitzGerald (1988) states that the amount of sand in these deltas is comparable in volume to that of the adjacent barrier islands. Therefore, on mixed-energy coasts, minor changes in volume of an ebb delta can drastically affect the supply of sand to the adjacent beaches. In comparison, on wave-dominated barrier coasts (e.g., Maryland, Outer Banks of North Carolina, northern New Jersey, Egypt's Nile Delta), ebb-tidal deltas are more rare and therefore represent a much smaller percentage of the overall coastal sand budget. As a result, volumetric changes in the ebb deltas have primarily local effects along the nearby beaches.

(f) Using data from tidal inlets throughout the United States, Walton and Adams (1976) showed that there is a direct correspondence between an inlet's tidal prism and the size of the ebb-tidal delta, with some variability caused by changes in wave energy. This research underscores how important it is that coastal managers thoroughly evaluate whether proposed structures might change the tidal prism, thereby changing the volume of the ebb-tide shoal and, in turn, affecting the sediment budget of nearby beaches.

(g) Ocean City, Maryland, is offered as an example of the effect of inlet formation on the adjacent coastline: the Ocean City Inlet was formed when Assateague Island was breached by the hurricane of 23 August 1933. The ebb-tide shoal has grown steadily since 1933 and now contains more than $6 \times 10^6$ m$^3$ of sand, located a mean distance of 1,200 m offshore. Since 1933, the growth of the ebb delta combined with trapping of sand updrift of the north jetty have starved the downdrift (southern) beaches, causing the shoreline along the northern few kilometers of Assateague Island to retreat at a rate of 11 m/year (data cited in Fitz-Gerald (1988)) (Figures IV-3-20 and IV-3-21).

(h) In contrast to Ocean City, the decrease in inlet tidal prisms along the East Friesian Islands has been beneficial to the barrier coast. Between 1650 and 1960, the area of the bays behind the island chain decreased by 80 percent, mostly due to reclamation of tidal flats and marshlands (FitzGerald, Penland, and Nummedal 1984). The reduction in area of the bays reduced tidal prisms, which led to smaller inlets, smaller ebb-tidal shoals, and longer barrier islands. Because of the reduced ebb discharge, less sediment was transported seaward. Waves moved ebb-tidal sands onshore, increasing the sediment supply to the barrier beaches.

(i) In many respects, ebb-tide deltas found at tidal inlets are similar to deltas formed at river mouths. The comparison is particularly applicable at rivers where the flow temporarily reverses during the flood stage of the tide. The main difference between the two settings is that river deltas grow over time, fed by fluvially supplied sediment. In contrast, at many tidal inlets, only limited sediment is supplied from the back bay, and the ebb deltas are largely composed of sand provided by longshore drift or reworked from the adjacent beaches. Under some circumstances, inlets and river mouths are in effect the same coastal form. During times of low river flow, the mouth assumes the characteristics of a tidal inlet with reversing tidal currents dominating sedimentation. During high river discharge, currents are unidirectional and fluvial sediment is deposited seaward of the mouth, where it can help feed the growth of a delta. Over time, a tidal inlet that connects a pond to the sea can be converted to a river mouth. This occurs when the back bay fills with fluvial sediment and organic matter. Eventually, rivers that formerly drained into the lagoon flow through channels to the inlet and discharge directly into the sea (for example, see the photograph of the Essex River Delta, Figure IV-3-11).

Figure IV-3-20. Ocean City Inlet, Maryland, September 1933. Ocean City is on Fenwick Island in the top center of the image, the Atlantic Ocean is to the right, and Assateague Island is on the bottom. This photograph was taken only one month after a hurricane breached the barrier island. Note waves breaking at the edge of a small ebb shoal. (Photograph from Beach Erosion Board archives)

(2) Flood-tidal deltas (shoals).

(a) A model of a typical flood-tide shoal is shown in Figure IV-3-18. Flood shoals with many of these features have been described in meso- and micro-tidal environments around the world (Germany (Nummedal and Penland 1981); Florida's east coast (Stauble et al. 1988); Florida's Gulf of Mexico coast (Wright, Sonu, and Kielhorn 1972); and New England (Boothroyd 1985)). The major components are:

Coastal Morphodynamics

Figure IV-3-21.  Ocean City Inlet, Maryland, 11 September 1995.  Before the 1933 hurricane breached the barrier island, Assateague and Fenwick Islands were joined and had a straight shoreline.  The former shoreline ran approximately along the seaward-most road (Photograph provided by USAE District, Baltimore)

- The *flood ramp*, which is a seaward-dipping sand surface dominated by flood-tidal currents. Sediment movement occurs in the form of sand waves (dunes), which migrate up the ramp.

- *Flood channels*, subtidal continuations of the flood ramp.

- The *ebb shield*, the high, landward margin of the tidal delta that helps divert ebb-tide currents around the shoal.

- *Ebb spits*, high areas mainly formed by ebb currents with some interaction with flood currents.

- *Spillover lobes*, linguoid, bar-like features formed by ebb-tidal current flow over low areas of the ebb shield.

(b) Although this model was originally derived from studies in mesotidal, mixed-energy conditions, it is also applicable to more wave-dominated, microtidal inlets (Boothroyd 1985). However, flood-tide shoals apparently are not formed in macrotidal shores.

(c) The high, central portion of a flood-tidal delta often extends some distance into an estuary or bay. This is the oldest portion of the delta and is usually vegetated by marsh plants. The marsh cap extends up to the elevation of the mean high water. The marsh expands aerially by growing out over the adjacent tidal flat. The highest, marsh-covered part of a flood shoal, or sometimes the entire shoal, is often identified on navigation charts as a "middle ground."

(3) Bed forms.

(a) Inlets contain a broad range of bed forms, from ripples due to oscillatory waves to dunes and antidunes caused by tidal currents. Water mass stratification can influence inlet flow and, therefore, bed form orientation. When a lagoon contains brackish water, salt wedge dynamics can occur, where the incoming flood flows under less dense bay water. Mixing between the two waters occurs along a horizontal density interface. During ebb tide, a buoyant planar jet forms at the seaward opening of the inlet similar to the effluent from rivers.

(b) Wright, Sonu, and Kielhorn (1972) described how density stratification affected flow at the Gulf of Mexico and Choctawhatchee Bay openings of East Pass, Florida. During flood tide, drogues and dye showed that the incoming salty Gulf of Mexico water met the brackish bay water at a sharp density front and then dove underneath (Figure IV-3-22). The drogues indicated that the sea water intruded at least 100 m beyond the front into Choctawhatchee Bay. This was the reason that bed forms within the channels displayed a flood orientation over time. This flood orientation can be seen in aerial photographs (Figure IV-3-5).

*f.   Inlet stability and migration.*[1]

(1) Background. Inlets migrate along the coast - or remain fixed in one location - because of complex interactions between tidal prism, wave energy, and sediment supply. Some researchers consider the littoral system to be the principal sediment source that influences the stability of inlets (Oertel 1988). Not all of the sediment in littoral transport is trapped at the mouths of inlets; at many locations, a large proportion may be bypassed by a variety of mechanisms. Inlet sediment bypassing is defined as "the transport of sand from the updrift side of the tidal inlet to the downdrift shoreline" (FitzGerald 1988). Bruun and Gerritsen (1959) described three mechanisms by which sand moves past tidal inlets:

- Wave-induced transport along the outer edge of the ebb delta (the terminal lobe).

- The transport of sand in channels by tidal currents.

- The migration of tidal channels and sandbars.

They noted that at many inlets, bypassing occurred through a combination of these mechanisms. As an extension of this earlier work, FitzGerald, Hubbard, and Nummedal (1978) proposed three models to explain inlet sediment bypassing along mixed-energy coasts. The models are illustrated in Figure IV-3-23 and are discussed below.

---

[1] Material in this section has been adapted from FitzGerald (1988).

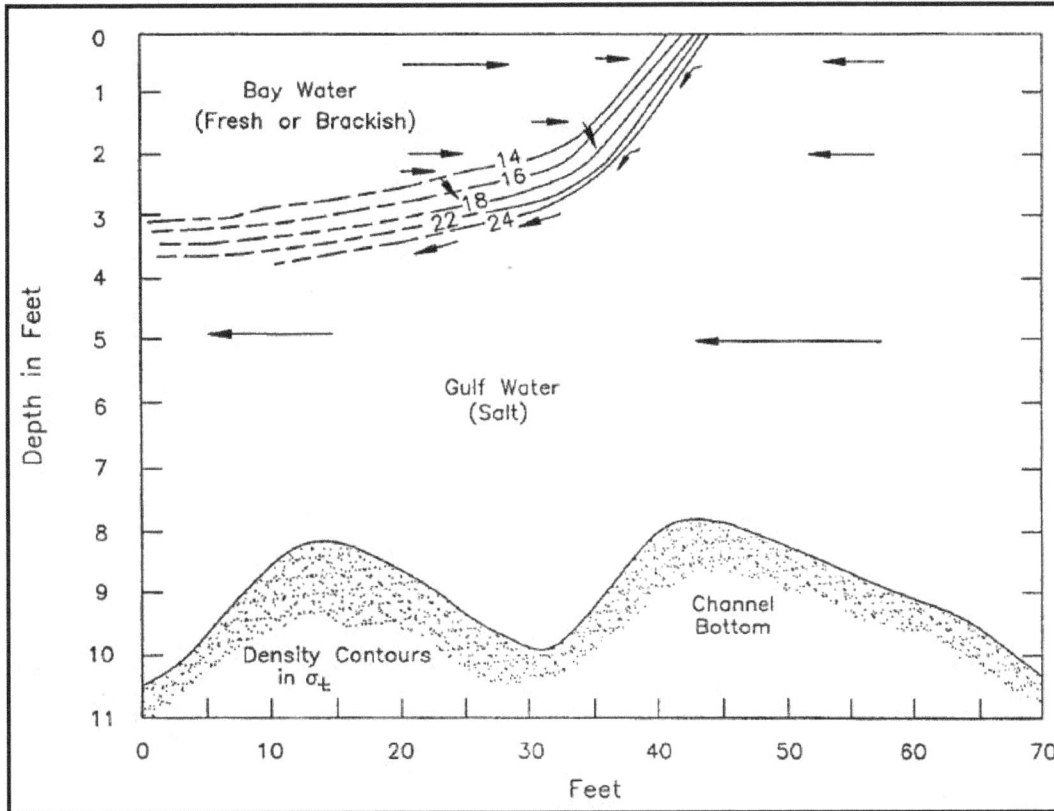

**Figure IV-3-22. At the bay opening of East Pass, Florida, stratified flow occurs during flood tide in Choctawhatchee Bay as a wedge of sea water dives underneath the lower density bay water (after Wright, Sonu, and Kielhorn (1972)). A similar phenomenon often occurs in estuaries**

(2) Inlet migration and spit breaching.

(a) The first model describes the tendency of many inlets to migrate downdrift and then abruptly shift their course by breaching a barrier spit. The migration occurs because sediment supplied by the longshore current causes the updrift barrier to grow (spit accretion). The growth occurs in the form of low, curved beach ridges, which weld to the end of the spit, often forming a bulbous-tipped spit known as a "drumstick." The ridges are often separated by low, marshy swales. As the inlet becomes narrower, the opposite (downdrift) shore erodes because tidal currents attempt to maintain an opening.

(b) In environments where the back bay is largely filled with marshes or where the barrier is close to the mainland, migration of the inlet causes an elongation of the tidal channel. Over time, the tidal flow between bay and ocean becomes more and more inefficient. Under these conditions, if a storm breaches the updrift barrier, the newly opened channel is a more direct and efficient pathway for tidal exchange. This new, shorter channel is likely to remain open while the older, longer route gradually closes. The breaching is most likely to occur across an area where the barrier has eroded or where some of the inner-ridge swales have remained low. The end result of spit accretion and breaching is the transfer of large quantities of sediment from one side of the inlet to the other. An example of this process is Kiawah River Inlet, South Carolina, whose migration between 1661 and 1978 was documented by FitzGerald, Hubbard, and Nummedal (1978). After a spit is breached and the old inlet closes, the former channel often becomes an elongated pond that parallels the coast.

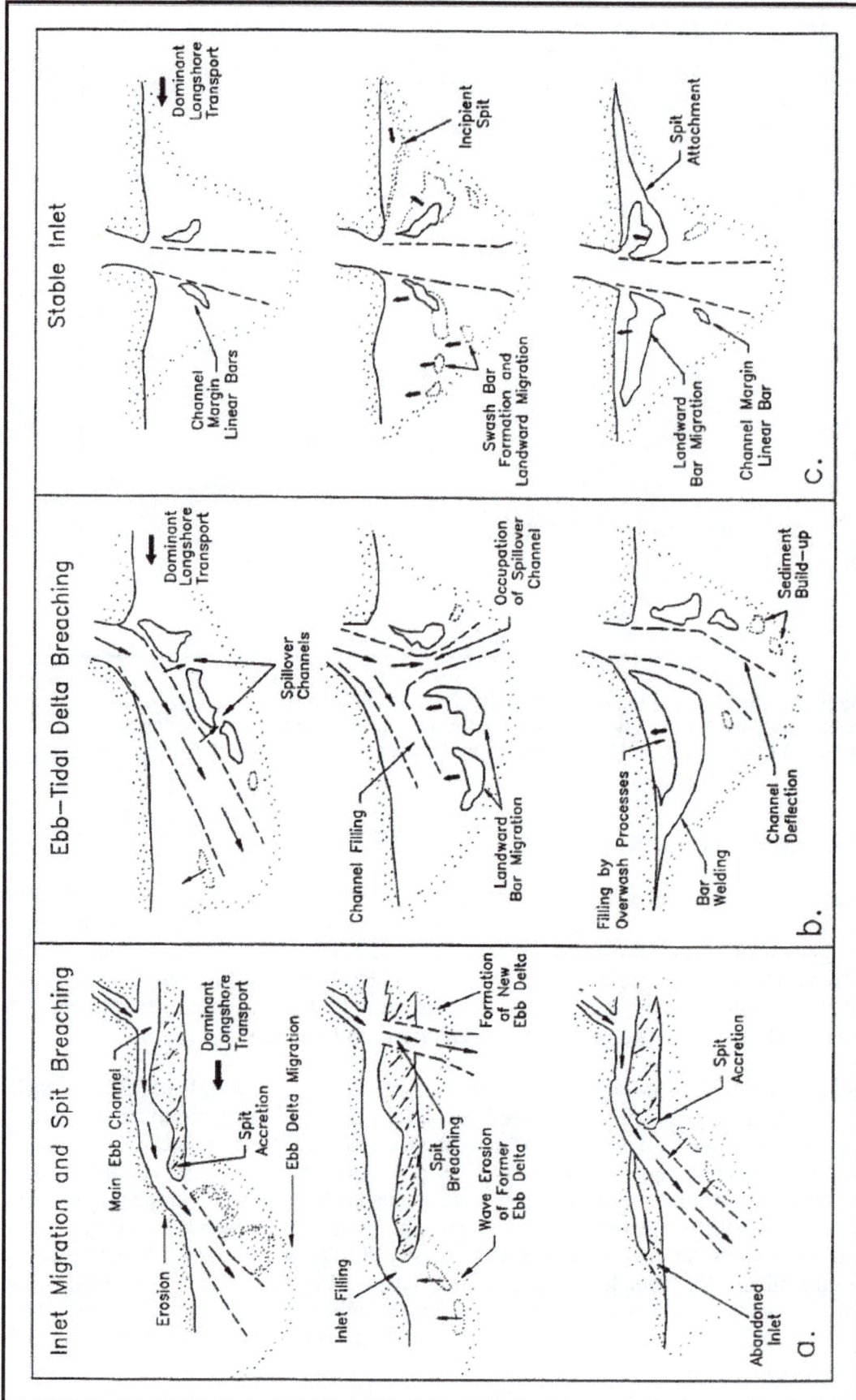

Figure IV-3-23. Three models of inlet behavior and sediment bypassing for mixed-energy coasts (from FitzGerald (1988))

(c) Several notes apply to the inlet migration model: First, not all inlets migrate. As discussed earlier, some inlets on microtidal shores are ephemeral, remaining open only a short time after a hurricane forces a breach through the barrier. If the normal tidal prism is small, these inlets are soon blocked by littoral drift. Short-lived inlets were documented along the Texas coast by Price and Parker (1979). The composition of the banks of the channel and the underlying geology are also critical factors. If an inlet abuts resistant sediment or bedrock, migration is restricted (for example, Hillsboro Inlet, on the Atlantic coast of Florida, is anchored by rock reefs). The gorge of deep inlets may be cut into resistant sediment, which also will restrict migration.

(d) Second, some inlets migrate updrift, against the direction of the predominate drift. Three mechanisms may account for updrift migration (Aubrey and Speer 1984):

- Attachment of swash bars to the inlet's downdrift shoreline.

- Breaching of the spit updrift of an inlet.

- Cutbank erosion of an inlet's updrift shoreline caused by back-bay tidal channels that approach the inlet throat obliquely.

(3) Ebb-tidal delta breaching.

(a) At some inlets, the position of the throat is stable, but the main ebb channel migrates over the ebb delta (Figure IV-3-23b). This pattern is sometimes seen at inlets that are naturally anchored by rock or have been stabilized by jetties. Sediment supplied by longshore drift accumulates on the updrift side of the ebb-tidal delta, which results in a deflection of the main ebb channel. The ebb channel continues to deflect until, in some cases, it flows parallel to the downdrift shore. This usually causes serious beach erosion. In this orientation, the channel is hydraulically inefficient, and the flow is likely to divert to a more direct seaward route through a spillover channel. Diversion of the flow can occur gradually over a period of months or can occur abruptly during a major storm. Eventually, most of the tidal exchange flows through the new channel, and the abandoned old channel fills with sand.

(b) Ebb delta breaching results in the bypassing of large amounts of sand because swash bars, which had formerly been updrift of the channel, become downdrift after the inlet occupies one of the spillover channels. Under the influence of waves, the swash bars migrate landward. The bars fill the abandoned channel and eventually weld to the downdrift beach.

(4) Stable inlet processes.

(a) These inlets have a stable throat position and a main ebb channel that does not migrate (Figure IV-3-23c). Sand bypassing occurs by means of large bar complexes that form on the ebb delta, migrate landward, and weld to the downdrift shoreline. The bar complexes are composed of swash bars that stack and merge as they migrate onshore. Swash bars are wave-built accumulations of sand that form on the ebb delta from sand that has been transported seaward in the main ebb channel. The swash bars move landward because of the dominance of landward flow across the swash platform. The reason for landward dominance of flow is that waves shoal and break over the terminal lobe (or bar) that forms along the seaward edge of the ebb delta. The bore from the breaking waves augments flood tidal currents but retards ebb currents.

(b) The amount of bypassing that actually occurs around a stable inlet depends upon the geometry of the ebb-tidal shoal, wave approach angle, and wave refraction around the shoal. Three sediment pathways can be identified:

- Some (or possibly much) of the longshore drift accumulates on the updrift side of the shoal in the form of a bar that projects out from the shore (Figure IV-3-23c). As the incipient spit grows, it merges with growing bar complexes near the ebb channel. Flood currents move some of the sand from the complexes into the ebb channel. Then, during ebb tide, currents flush the sand out of the channel onto the delta (both the updrift and downdrift sides), where it is available to feed the growth of new swash bars.

- Depending on the angle of wave approach, longshore currents flow around the ebb shoal from the updrift to the downdrift side. Some of the drift is able to move past the ebb channel, where it either continues moving along the coast or accumulates on the downdrift side of the ebb shoal. The sediment probably moves as large bed forms (Figure IV-3-4).

- Wave refraction around some ebb shoals causes a local reversal of longshore current direction along the downdrift shore. During this time, presumably, little sediment is able to escape the confines of the ebb-tidal shoal.

(5) Extension of bypassing models to other environments. The inlet migration models described above were originally based on moderate- to high-energy shores. However, research along the Florida Panhandle suggests that the models may be applicable to much lower energy environments than the original authors had anticipated. For example, between 1870 and 1990, the behavior of East Pass inlet, located in the low-wave-energy Florida Panhandle, followed all three models at various times (Figure IV-3-24; Morang (1992b)). It would be valuable to conduct inlet studies around the world to further refine the models and evaluate their applicability to different shores.

g. *Inlet response to jetty construction and other engineering activities.*

(1) Introduction. Typically, jetties are built to stabilize a migrating inlet, to protect a navigation channel from waves, or to reduce the amount of dredging required to maintain a specified channel depth. However, jetties can profoundly affect bypassing and other processes at inlets. Some of these effects can be predicted during the design phase of a project. Unfortunately, unanticipated geological conditions often arise, which lead to problems such as increased shoaling or changes in the tidal prism. Several classes of man-made activities affect inlets:

(a) Jetties stabilize inlets and prevent them from migrating.

(b) Jetties can block littoral drift.

(c) Walls or revetments can change the cross section of an inlet.

(d) Dredging can enlarge the cross section of a gorge.

(e) Dam construction and freshwater diversion reduce fluvial input.

(f) Weir sections (low portions of a jetty) allow sediment to pass into an inlet, where it can accumulate in a deposition basin and be bypassed.

(g) Landfilling and development in estuaries and bays can reduce tidal prism.

(2) Technical literature. Many reports have documented the effects of jetties on littoral sediment transport. Early works are cited in Barwis (1976). Dean (1988) discussed the response of modified Florida inlets,

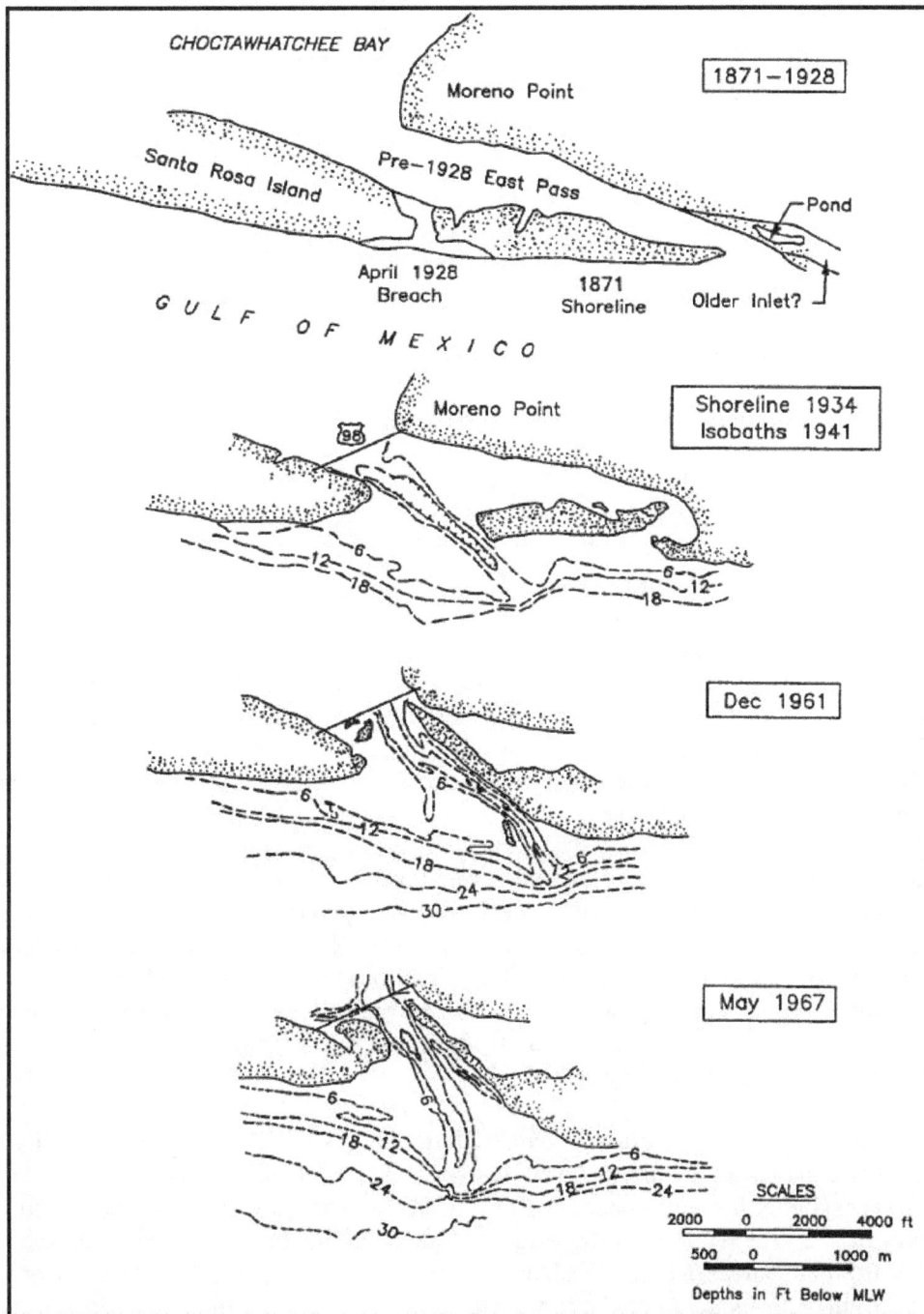

**Figure IV-3-24. Spit breaching and inlet migration at East Pass, Florida (from Morang (1992a))**

and many other case studies are reviewed in Aubrey and Weishar (1988). Examples of monitoring studies that assess the effects of jetties include:

(a) Shinnecock Inlet, Long Island, New York (Morang, 1999; Pratt and Stauble 2001; Williams, Morang, and Lillycrop 1998)

(b) Ocean City Inlet, Maryland (Bass et al. 1994).

(c) Little River Inlet, North and South Carolina (Chasten 1992, Chasten and Seabergh 1992).

(d) Murrells Inlet, South Carolina (Douglass 1987).

(e) Manasquan Inlet, New Jersey (Bruno, Yavary, and Herrington (1998).

(f) St. Marys Entrance, Florida and Georgia (Kraus, Gorman, and Pope 1994, 1995).

(g) East Pass, Florida (Morang 1992a).

(h) Port Mansfield Channel, Texas (Kieslich 1977).

(3) Weirs and other structures and their effects on sediment movement in and around inlets are discussed in Part V-6.

*h.  Case study of inlet formation and growth:  Shinnecock Inlet, Long Island, New York.*[1]

(1) Background and inlet breaching (1938).

(a) Shinnecock Inlet is located on the south shore of Long Island, 136 km east of New York Harbor. It is the easternmost of six inlets that cut the south shore barrier islands and allow boat passage between the Atlantic Ocean and the coastal bays. The present inlet was breached during the Great New England Hurricane of 21 September 1938. Inlets had periodically existed in this area before, but immediately before the 1938 hurricane, the barrier was intact and there was a paved road along the beach (Nersesian and Bocamazo 1992) (Figure IV-3-25).

(b) The first aerial photographs of the new inlet were taken by the U.S. Army Air Corps three days after the hurricane (Figure IV-3-26). The seas had calmed, but the large number of overwash deposits attest to the violence of the storm. All the open breaches in this area trended left of perpendicular to the shoreline trend. Drift was to the east (opposite to the normal prevailing direction) because spits had grown from west-to-east across the mouths of the new openings. Two months after the hurricane (29 November 1938) an oval ebb-tide delta had already formed in the Atlantic Ocean, and there was a small shoal in Shinnecock Bay, showing the initial development of a flood shoal (Figure IV-3-27).

(c) There are no eye-witness accounts of exactly how this inlet was cut. One possibility is that storm waves from the ocean cut across the barrier. The overwash fans to the left and right of the inlet support this option. The other possibility is that Shinnecock Bay elevated due to rainwater and runoff. This is supported by a water level of 2.2 (7.2 ft) above mlw estimated at the south end of the Shinnecock canal, across Shinnecock Bay from the barrier island (U.S. Army Corps of Engineers 1958)). At a low or narrow place in the barrier, a torrent of bay water might have burst through the dunes and run out to sea, scouring a channel that later widened and became the inlet shown in the photograph. This latter hypothesis seems most likely because, only two months after the storm, a prominent ebb shoal already existed, while there was only minimal deposition in Shinnecock Bay. The ebb shoal likely consisted of sand eroded from the barrier.

(2) Semi-stabilized inlet (1947). After 1938, the inlet migrated steadily to the west. In 1939-40, Suffolk County erected timber pile bulkhead and short groins on the west side of the channel to prevent westward migration (Nersesian and Bocamazo 1992). However, a 1947 photograph shows that the inlet had moved

---

[1]  Condensed from Morang (1999).

Figure IV-3-25.  Shinnecock Bay, Long Island, New York, 30 June 1938.  Photograph taken before the Great New England Hurricane of 21 September cut the barrier island in several locations.  The present Shinnecock Inlet was cut where a channel crosses from the bay to the road.  The channel may be a remnant of an artificial cut that was dug by the Shinnecock and Peconic Canal Company in 1896.  Shoal areas in the bay near the barrier attest to former inlets that had been open long enough to form flood shoals.  North is to the top, the water body in the center of the image is Shinnecock Bay, and the Atlantic Ocean is on the bottom.  The image is part of a mosaic prepared by the Beach Erosion Board

east, leaving the bulkhead behind a sand beach (Figure IV-3-28).  A spit extended out to sea from the west shore, suggesting that longshore transport, at least temporarily, had been to the east (possibly a local reversal). Some small shoals existed in the mouth of the inlet, but sediment transport was largely directed into the bay after 1938.  The ebb shoal was a flat oval that hugged the shore.  The shoreline in this area trends about 65° (approx NE-SW).  The 1947 channel extended out to sea at about 120°, more eastward than perpendicular to the shore.  In the following years, the inlet rotated clockwise, until by 1951, the channel was pushing up against the bulkhead and the flood shoal deposition was occurring east of its earlier location.

(3) Stabilized inlet (1996).  East and west jetties were built in 1952-54 by Suffolk County and the State of New York and rebuilt in 1992-93 by the Corps of Engineers.  A photograph from 24 October 1996 (Figure IV-3-29) shows that the channel inlet was aligned north-south, about 40° clockwise from the orientation in 1947.  The jetties were 240 m (800 ft) apart.  The updrift beach had advanced almost to the tip of the east jetty.  During the 1970's and 1980's, the bay shoreline east of the inlet eroded about 500 ft (150 m) due to scour from the east ebb channel.  To the west, the shore has suffered chronic erosion, requiring frequent emergency repair by the highway department to prevent the road from being breached.  The ebb shoal (only partially shown in this photograph) is an unsymmetric oval that attaches to the downdrift beach about 1,200 m west of the west jetty.

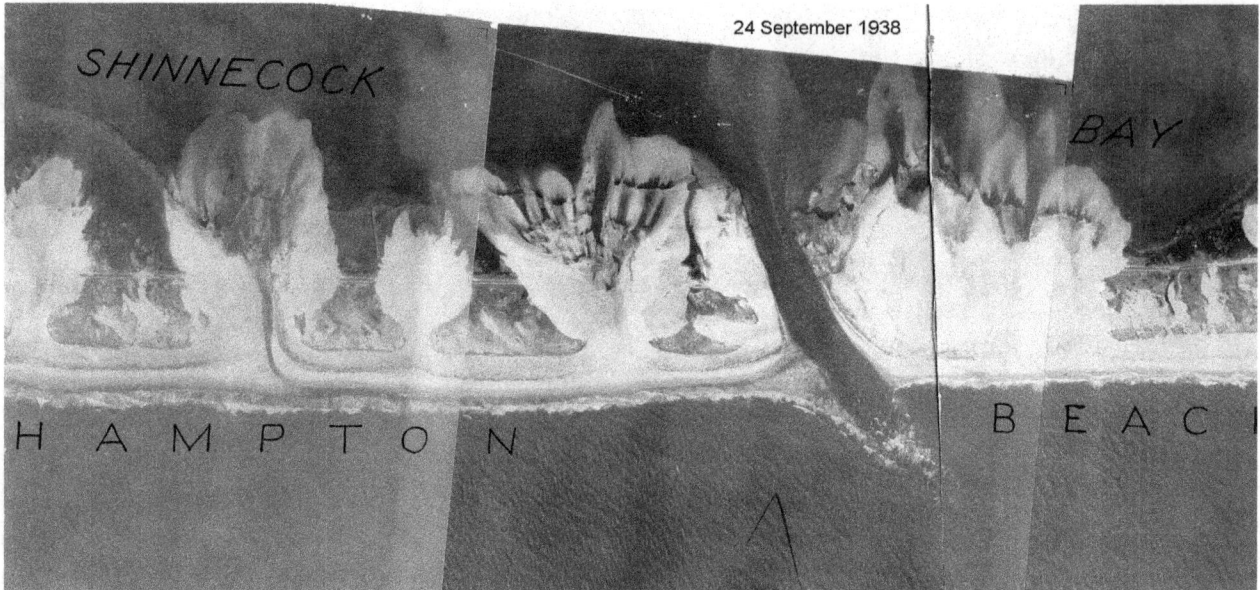

**Figure IV-3-26. Shinnecock Inlet, Long Island, New York, 24 September 1938. Taken three days after Great New England Hurricane breached the barrier, this image shows the new Shinnecock Inlet and many overwash fans along the adjacent shore. Immediately after the storm, all the inlets along this stretch of the coast trended left of the shore-perpendicular. (Photograph mosaic from Beach Erosion Board archives)**

(4) Ebb shoal growth. The ebb shoal at Shinnecock Inlet has continued to grow in volume since the inlet was formed. Figure IV-3-30 shows that in 1998, the shoal contained about 6,000,000 m$^3$ (8,000,000 yd$^3$) of sand. It appears that the shoal is still growing, but the latest three surveys were spaced only 1 year apart (1996, 1997, and 1998), so another survey will be needed in several years to verify this conclusion. However, present evidence indicates that the shoal is a sediment sink, although some of the littoral drift is probably bypassing.

## IV-4-5. Morphodynamics and Shoreface Processes of Clastic Sediment Shores

*a. Overview.*

(1) Introduction. This section discusses morphodynamics - the interaction of physical processes and geomorphic response - of clastic sediment shores. The topic covers beach features larger than a meter (e.g., cusps and bars) on time scales of minutes to months. Details on grain-to-grain interactions, the initiation of sediment motion, and high frequency processes are not included. A principle guiding this section is that the overall shape of beaches and the morphology of the shoreface are largely a result of oscillatory (gravity) waves, although tide range, sediment supply, and overall geological setting impose limits. We introduce basic relationships and formulas, but the text is essentially descriptive. Waves are discussed in detail in Part II of the *Coastal Engineering Manual*, while sediment properties are covered in Part III-1.

(2) Literature. Beaches and sediment movement along the shore have been subjects of popular and scientific interest for over a century. A few of the many textbooks that cover these topics include Carter (1988), Davis (1985), Davis and Ethington (1976), Greenwood and Davis (1984), Komar (1998), and Zenkovich (1967). Small-amplitude (Airy) and higher-order wave mechanics are covered in EM 1110-2-1502; more detailed treatments are in Kinsman (1965), Horikawa (1988), and Le Méhauté (1976). Interpreting and applying wave and water level data are covered in EM 1110-2-1414.

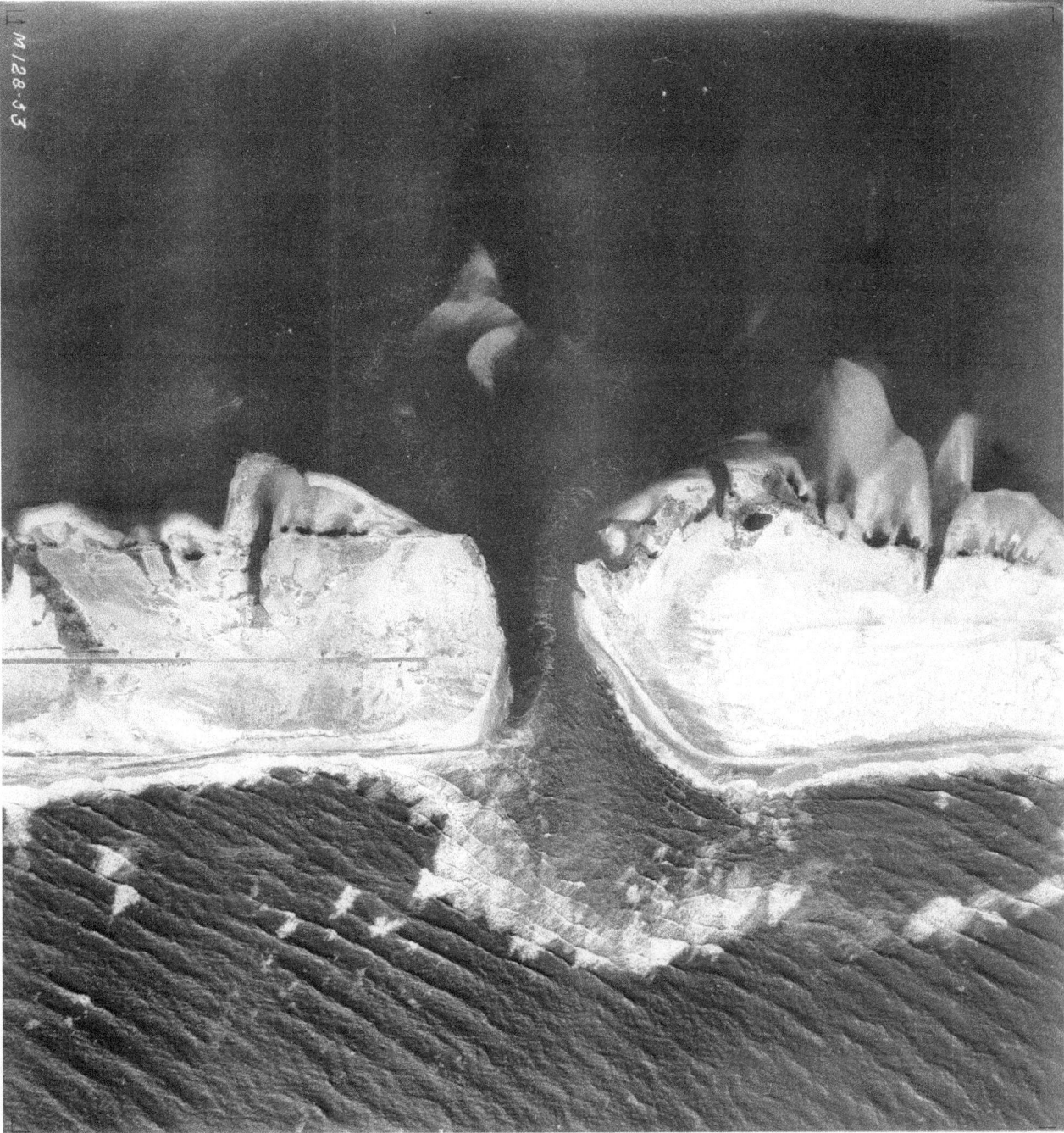

Figure IV-3-27.  Shinnecock Inlet, New York, 29 November 1938, photographed only 2 months after the barrier island was breached during the Great New England Hurricane of 21 September 1938.  North is to the top of the image.  The body of water to the top is Shinnecock Bay, and the Atlantic Ocean is on the bottom.  The inlet already displays an oval ebb-tidal shoal.  A small shoal in Shinnecock Bay indicated development of a flood shoal.  Washover fans, remnants of the storm, can be seen on both sides of the inlet.  (Photograph from Beach Erosion Board archives)

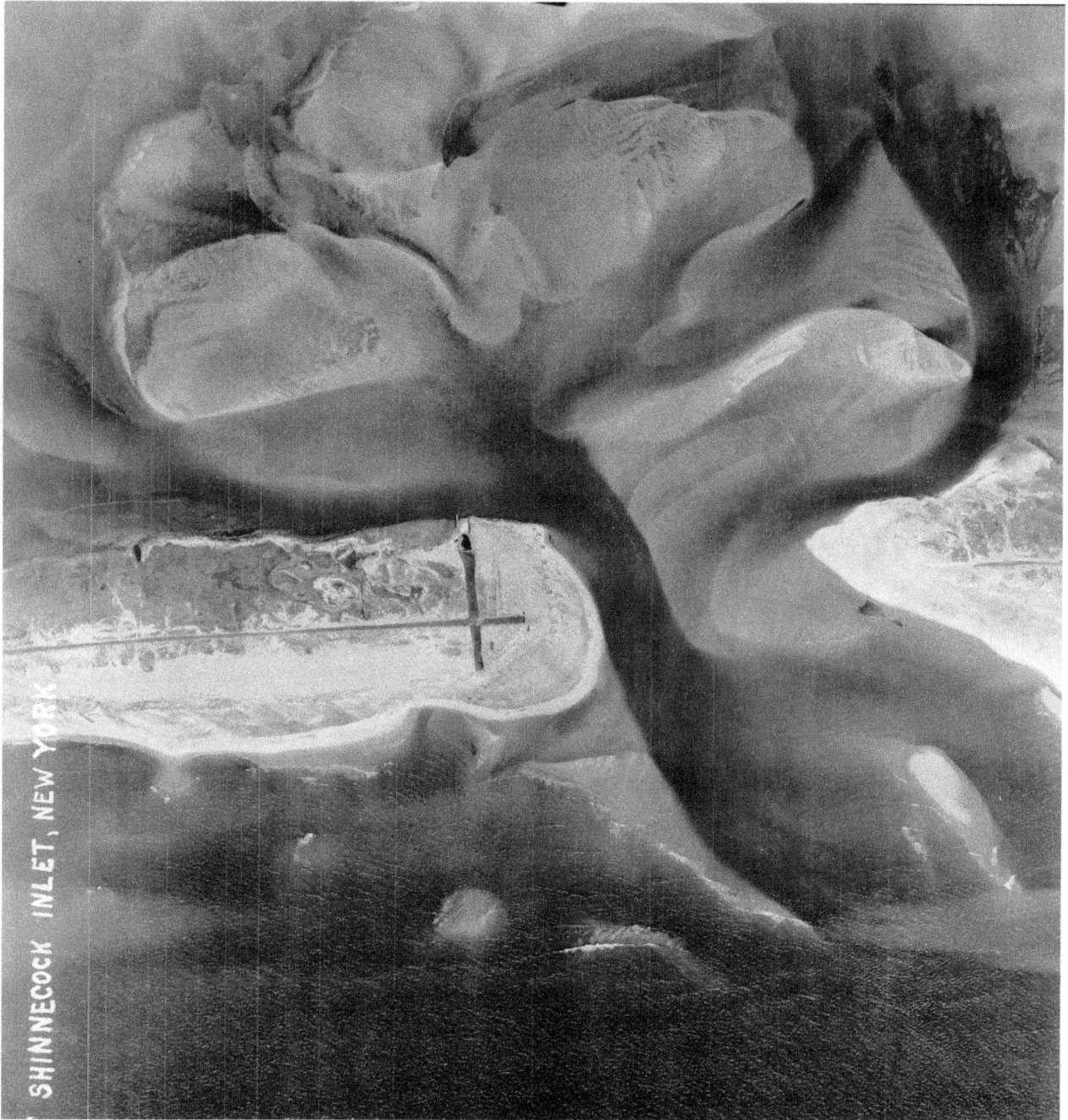

Figure IV-3-28.  Shinnecock Inlet, New York, 1 April 1947.  The inlet was much wider than in 1938, and a large flood shoal had accumulated.  Some small shoals existed in the mouth of the inlet, but sediment transport had been largely directed into the bay since 1938.  (Photograph from Beach Erosion Board archives)

(3)  Significance of clastic coasts.  It is important to examine and understand how clastic shores respond to changes in wave climate, sediment supply, and engineering activities for economic and management reasons:

(a)  Many people throughout the world live on or near beaches.  Beaches are popular recreation areas and of vital economic importance to many states (Houston 1995, 1996a, 1996b).

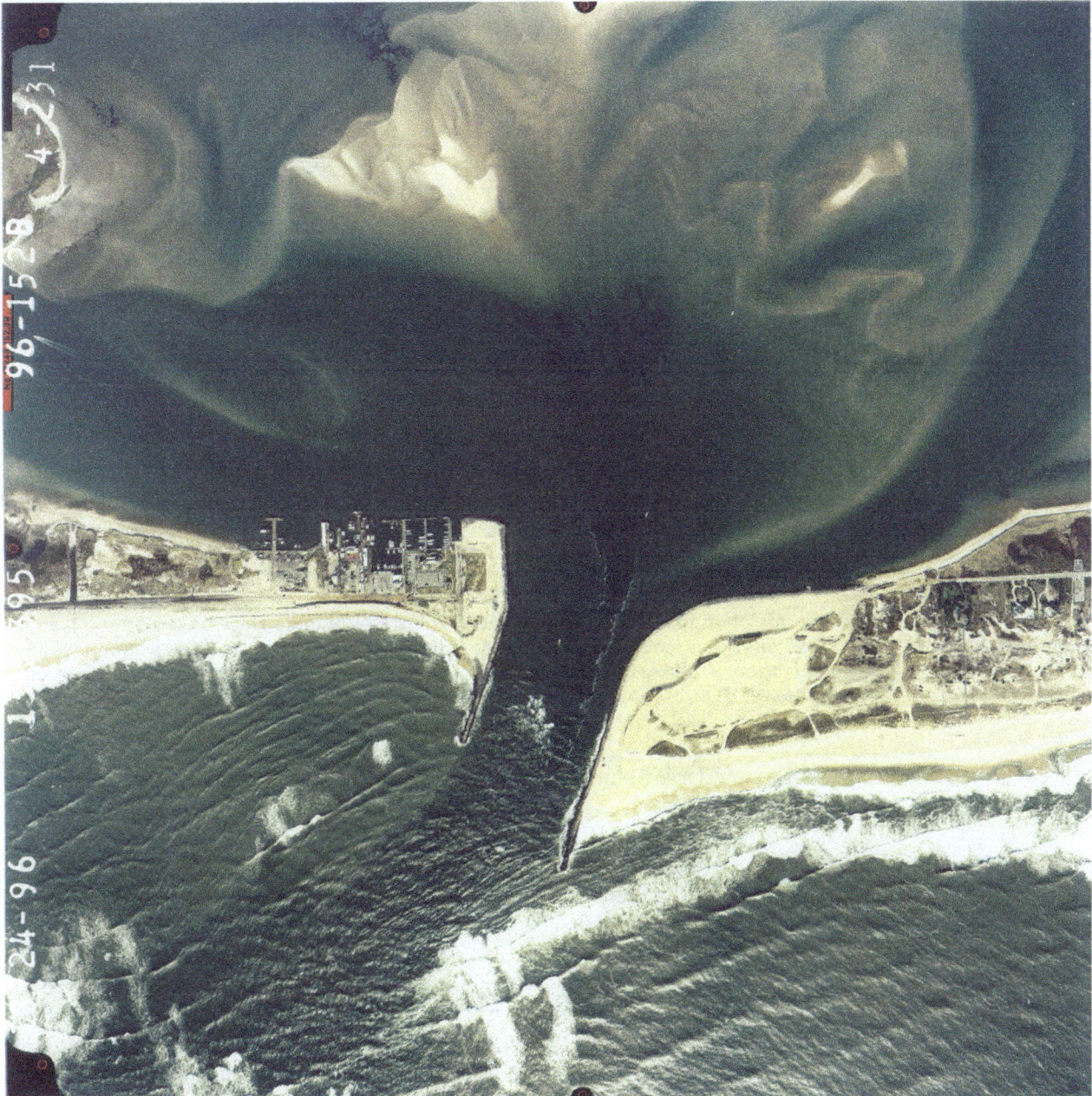

Figure IV-3-29. Shinnecock Inlet, New York, 24 October 1996. The inlet has been stabilized with stone jetties 800 ft apart. The beach east of the inlet has advanced to the end of the east jetty. To the west, the shore has suffered chronic erosion, requiring frequent emergency repair by the highway department to prevent the road from being breached. The ebb shoal (only partially shown in this photograph) is an unsymmetric oval that attaches to the downdrift beach about 1,200 m west of the west jetty (Photograph courtesy of USAED, New York)

(b) Beaches are critical buffer zones protecting wetlands and coastal plains from wave attack.

(c) Beaches are habitat or nesting ground for many animal species, some of which are endangered (e.g., turtles).

## SHINNECOCK INLET, NEW YORK
### Ebb-Shoal Volume Changes

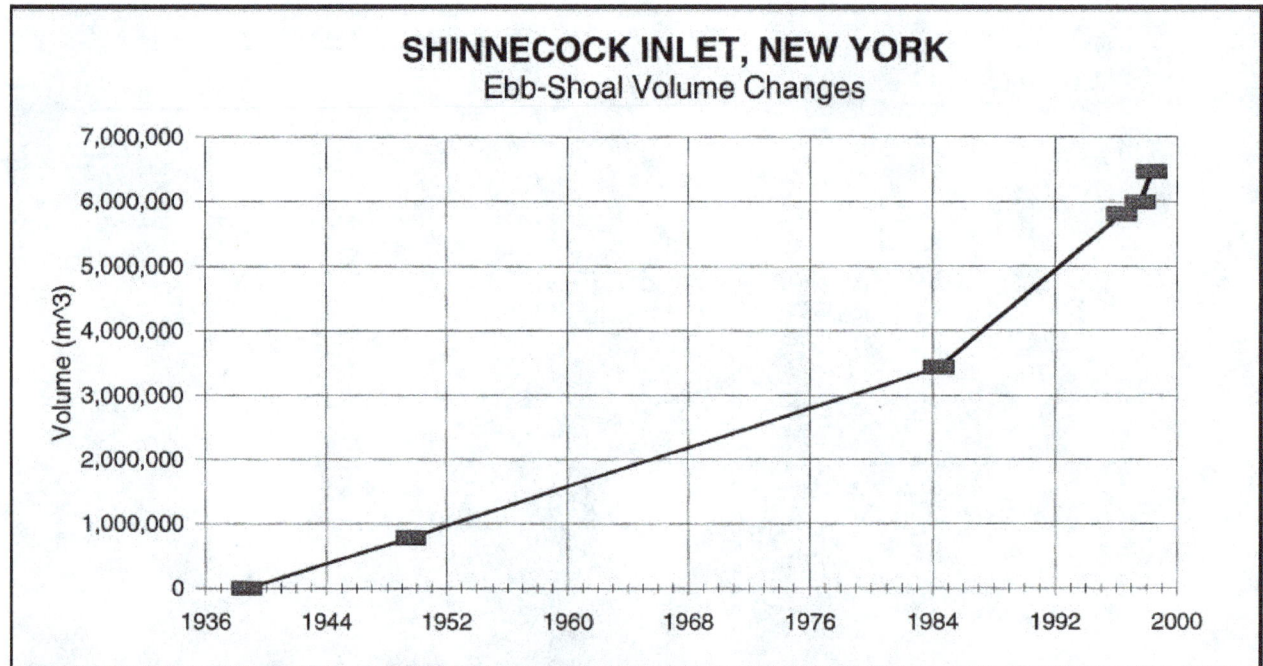

**Figure IV-3-30. Ebb-shoal volume changes, Shinnecock Inlet, New York. Volumes were based on subtracting the 1933 U.S. Coast and Geodetic Survey data (pre-inlet) from more recent surveys: 1949, 1985, 1996, 1997, and 1998. The ebb shoal is still growing, indicating that it is a sediment trap, although some of the littoral drift probably bypasses. Hydrographic data courtesy of U.S. Army Engineer District, New York**

(d) Much engineering effort and large amounts of funding are expended on planning and conducting beach renourishment.

(e) Sediment supply and, therefore, beach stability, are often adversely affected by the construction of navigation and shore-protection structures.

(f) Sand is a valuable and increasingly rare mineral resource in most of the coastal United States.

(4) Range of coastal environments. Around the world, coasts vary greatly in steepness, sediment composition, and morphology. The most dynamic shores may well be those composed of unconsolidated clastic sediment because they change their form and state rapidly. Clastic coasts are part of a geologic continuum that extends from consolidated (rocky) to loose clastic to cohesive material (Figure IV-3-31). Waves are the primary mechanisms that shape the morphology and move sediment, but geological setting imposes overall constraints by controlling sediment supply and underlying rock or sediment type. For example, waves have little effect on rocky cliffs; erosion does occur over years, but the response time is so long that many rocky shores can be treated as being geologically controlled. At the other end of the continuum, cohesive shores respond very differently to wave action because of the electrochemical nature of the sediment (see Part III-5).

  *b. Tide range and overall beach morphology.*

Most studies of beach morphology and processes have concentrated on microtidal (< 1 m) or low-mesotidal coasts (1-2 m). To date, many details concerning the processes that shape high-meso- and macrotidal beaches (tide range > 2 m) are still unknown. Based on a review of the literature, Short (1991) concluded that

# GEOLOGIC SETTING

## MORPHODYNAMIC CONTROL

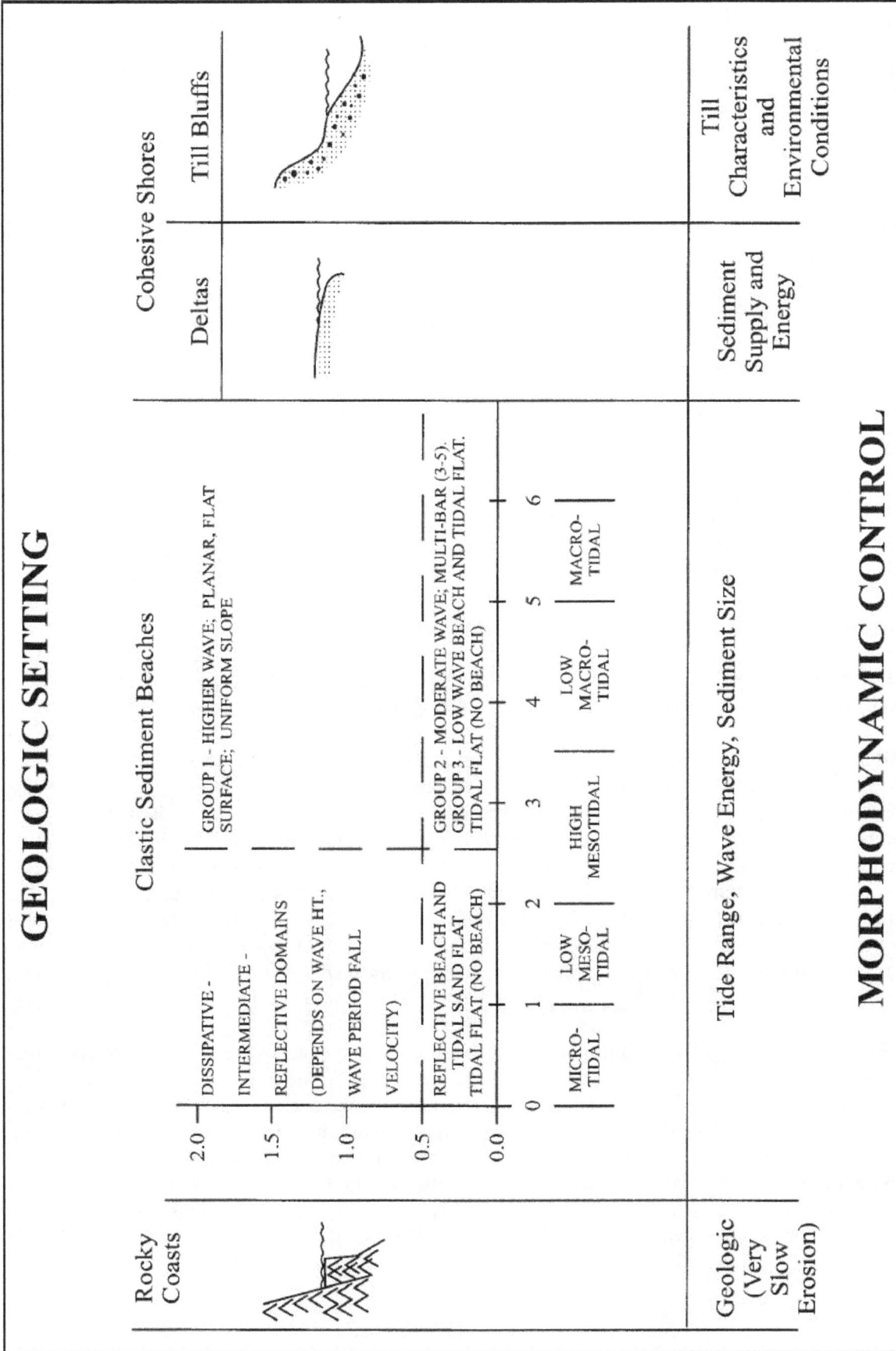

Figure IV-3-31. Summary of factors controlling morphodynamics along a range of coastal environments. Clastic shore processes are detailed in Figure IV-3-26 and discussed in the text

wave-dominated beaches where tide range is greater than about 2 m behave differently than their lower-tide counterparts. Short underscored that high-tide beaches are also molded by wave and sediment interactions. The difference is the increasing impact of tidal range on wave dynamics, shoreface morphodynamics, and shoreline mobility. Short developed a tentative grouping of various beach types (Figure IV-3-32). Discussion of the various shoreface morphologies follows.

   *c.  High tidal range (> 2 m) beach morphodynamics.*

   (1) Review. Based on a review of research on macrotidal beaches, Short (1991) summarized several points regarding their morphology:

   (a) They are widespread globally, occurring in both sea and swell environments.

   (b) Incident waves dominate the intertidal zone.

   (c) Low-frequency (infragravity) standing waves may be present and may be responsible for multiple bars.

   (d) The intertidal zone can be segregated into a coarser, steeper, wave-dominated high tide zone, an intermediate zone of finer sediment and decreasing gradient, and a low-gradient, low-tide zone. The highest zone is dominated by breaking waves, the lower two by shoaling waves.

   (e) The cellular rip circulation and rhythmic topography that are so characteristic of microtidal beaches have not been reported for beaches with tide range greater than 3 m.

   (2) Macrotidal beach groups. Using published studies and field data from Australia, Short (1991) divided macrotidal beaches into three groups based on gradient, topography, and relative sea-swell energy:

   (a) GROUP 1 - High wave, planar, uniform slope. Beaches exposed to persistently high waves ($H_b > 0.5$ m) display a planar, flat, uniform surface. Shorefaces are steep, ranging from 1 to 3 deg, and have a flat surface without ripples, bed forms, or bars. The upper high tide beach is often relatively steep and cuspid and contains the coarsest sediment of the system. On both sand and gravel beaches, the high tide, upper foreshore zone is exposed to the highest waves. Plunging and surging breakers produce asymmetric swash flows, which maintain the coarse sediment and steep gradient. Further seaward, wave shoaling becomes a more important factor than wave breaking because waves are attenuated at low tide (due to shallower water and greater friction). Tidal currents also increase in dominance seaward. Wright (1981) found that tidal currents left no bed forms visible at low tide but were an important factor in longshore sediment transport.

   (b) GROUP 2 - Moderate wave, multi-bar. Multi-bar, macrotidal beaches are formed in fetch-limited environments with high tide range and abundant fine sand (King 1972b). The common characteristic of these beaches is a relatively uniform 0.5- to 0.6-deg intertidal gradient and the occurrence of multiple bars (two to five sets) between msl and mlw (Short 1991). Bar amplitude is usually below 1 m and spacing ranges from 50 to 150 m, with spacing increasing offshore. Field observations indicate that the bars are formed by a wave mechanism, particularly during low wave, post-storm conditions. The bars appear to build up onsite rather than migrate into position. These multi-bar beaches probably cause dissipative conditions during most wave regimes, possibly resulting in the development of infragravity standing waves. This would account for the spacing of the bars; however, this hypothesis has not been tested with rigorous field measurements (Short 1991).

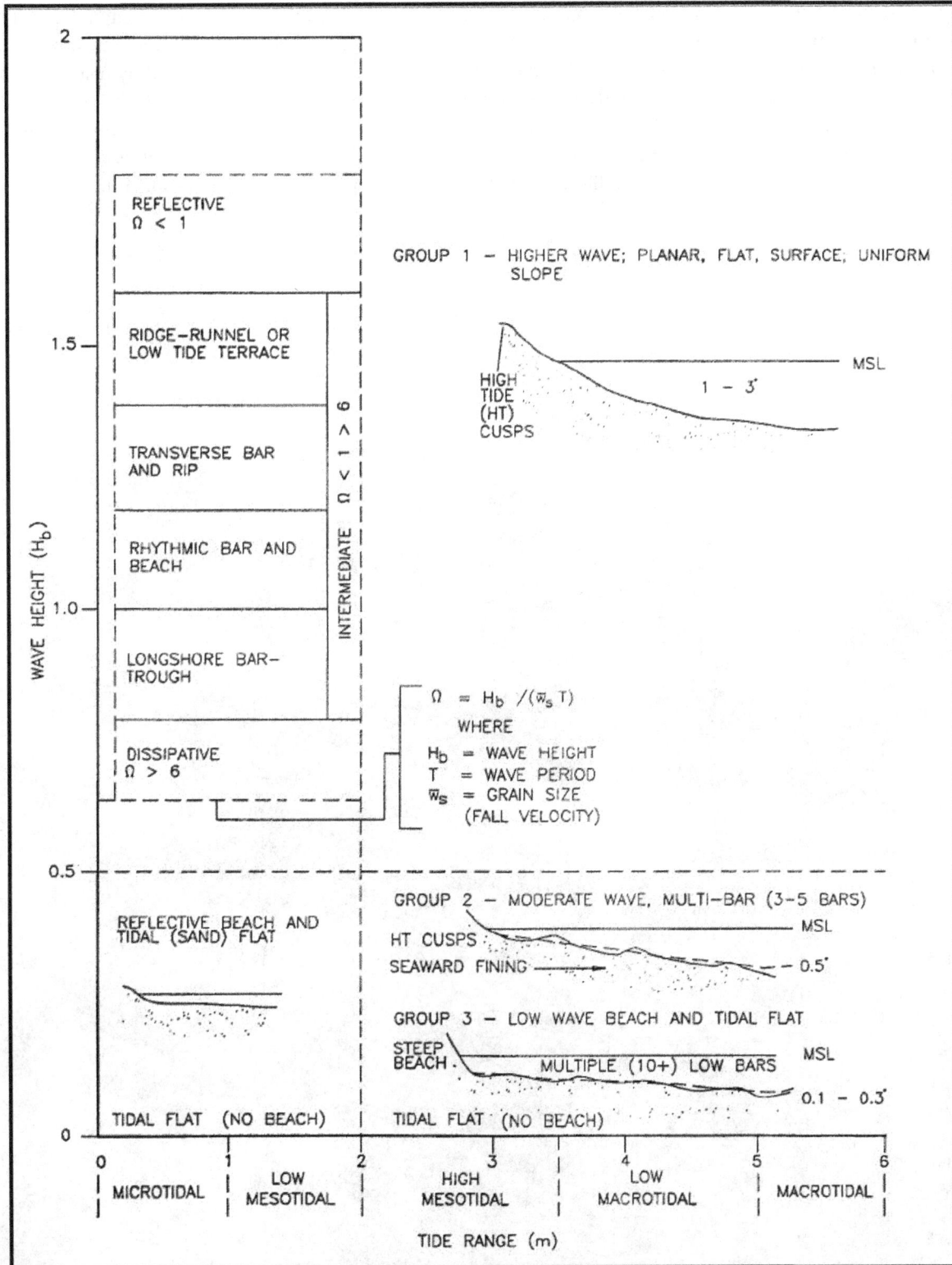

Figure IV-3-32. Micro- to macrotidal beach and tidal flat systems (adapted from Short (1991)). Dimensionless parameter Ω discussed in the text

(c) GROUP 3 - Low wave beach and tidal flat. As wave energy decreases, macrotidal beaches eventually grade into tide-dominated tidal flats. Between the two regimes, there is a transition stage that contains elements of both morphologies. These beach-tidal flat systems are usually characterized by a steep, coarse-grained reflective beach (no cusps usually present) which grades abruptly at some depth below msl into a fine-grained, very low gradient (0.1 deg), rippled tidal flat (Figure IV-3-33). The tidal flat may be uniform or may contain low, multiple bars. Beach-tidal flat shores are found in low-energy environments that are only infrequently exposed to wave attack, but the energy must be sufficient to produce the morphologic zonation.

Figure IV-3-33.   View west from Tarskavaig, Isle of Skye, Scotland, August 1983.  Photograph taken at low tide.  At this macrotidal beach, facing the Sea of the Hebrides, the upper shoreface consists of coarse cobble, while the low foreshore is low gradient and almost featureless

(3) Spatial and temporal variations. Beaches on macro-tidal coasts vary morphologically as important environmental parameters change. Short (1991) cites one setting where the shoreface varies from high-energy, uniform steep beach (Group 1) to beach-tidal flat (Group 3) within 2 km. He suggests that the changes in morphology are due to variations in wave energy: as energy changes alongshore, important thresholds are crossed which result in different ratios of wave versus tide domination. In addition, there may be temporal variations throughout the lunar cycle. As tide range varies during the month, the transitions where one morphologic group merges into another may migrate cyclically along the coast. More field studies are needed to document this phenomenon.

(4) Summary. On tideless beaches, morphology is determined by waves and sediment character. On microtidal beaches, waves still dominate the morphodynamics, but tide exerts a greater influence. As tide range increases beyond 2-3 m, the shape of beaches becomes a function of waves coupled with tides. On the higher tide coasts, as water depth changes rapidly throughout the day, the shoreline and zone of wave breaking move horizontally across the foreshore and tidal currents move considerable sediment.

*d. Morphodynamics of micro- and low-mesotidal coasts.*

(1) Morphodynamic variability of microtidal beaches and surf zones. Based on field experiments in Australia, Wright and Short (1984) have presented a model of shoreface morphology as a function of wave parameters and sediment grain size. This model is a subset of Figure IV-3-31 that occupies the zone where tide range is between 0 and 2 m and $H_b$ (breaker height) is greater than about 0.5 m.

(a) Wright and Short (1984) determined that the morphodynamic state of sandy beaches could be classified on the basis of assemblages of depositional forms and the signatures of associated hydrodynamic processes. They identified two end members of the morphodynamic continuum:

- Fully dissipative.

- Highly reflective.

Between the extremes were four intermediate states, each of which possessed both reflective and dissipative elements (Figure IV-3-32).

(b) The most apparent differences between the beach states are morphological, but distinct process signatures, representing the relative velocities of different modes of fluid motion, accompany the characteristic morphology. As stated by Wright and Short (1984):

Although wind-generated waves are the main source of the energy which drives beach changes, the complex processes, which operate in natural surf zones and involve various combinations of dissipation and reflection, can lead to the transfer of incident wave energy to other modes of fluid motion, some of which may become dominant over the waves themselves.

(c) Wright and Short grouped fluid motion into four categories (Table IV-3-1):

- Oscillatory flows.

- Oscillatory or quasi-oscillatory flows.

- Net circulations.

- Non-wave-generated currents.

**Table IV-3-1**
**Modes of Fluid Motion Affecting Clastic Shorelines**

| Modes | Notes | Frequencies of flows | Examples |
|---|---|---|---|
| Oscillatory | Corresponds directly to incident waves | Frequency band of deep-water incident waves | Sediment-agitating oscillations |
| Oscillatory or quasi-oscillatory | Shore-normal oriented standing and edge waves | Wide range of frequencies | Trapped edge waves, "leaky" mode standing waves |
| Net circulations | Generated by wave energy dissipation | Minutes to days | Longshore currents, rip currents, rip feeder currents |
| Non-wave-generated currents | Generated by tides and wind shear | Minutes to hours (?) | Tidal currents |

(Based on Wright and Short (1984))

(d) From repeated observations and surveys of beaches, Wright and Short (1984) concluded that beach state is clearly a function of breaker height and period and sediment size. Over time, a given beach tends to exhibit a *modal* or most frequent recurrent state, which depends on environmental conditions. Variations in shoreline position and profile are associated with temporal variations of beach state around the modal state. Wright and Short found that a dimensionless parameter $\Omega$ could be used to describe the modal state of the beach:

$$\Omega = \frac{H_b}{\overline{w_s}T} \tag{IV-3-3}$$

where $H_b$ is breaker height, $\overline{w_s}$ is sediment fall velocity, and $T$ is wave period. For the following values of $\Omega$ (Figure IV-3-32):

- $\Omega$ about 1:  defines the reflective/intermediate threshold

- $1 < \Omega < 6$:  intermediate beaches

- $\Omega \cong 6$:  marks the threshold between intermediate and dissipative conditions

(e) Beaches take time to adjust their state, and a change of $\Omega$ across a threshold boundary does not immediately result in a transformation from reflective to intermediate or from intermediate to dissipative. On the Pacific coasts of Australia and the United States, storms can cause a shift of beach state from reflective or intermediate to dissipative in a few days because the energy is high. The return to reflective conditions under low energy may require weeks or months or longer (the sequence of beach recovery is illustrated in stages *a* through *f* in Figure IV-3-34). In environments where the dominant variation in wave energy occurs on an annual cycle (e.g., high storm waves in winter and low swell in summer), the full range from a dissipative winter profile to a reflective summer profile may be expected.

(f) Wright and Short (1984) concluded that large temporal variations in $\Omega$ are accompanied by large changes in state. However, when the variations in $\Omega$ take place in the domains of $\Omega < 1$ or $\Omega > 6$, no corresponding changes in *state* result. Intermediate beaches, where $\Omega$ is between 1 and 6, are spatially and temporally the most dynamic. They can undergo rapid changes as wave height fluctuates, causing reversals in onshore/offshore and alongshore sediment transport.

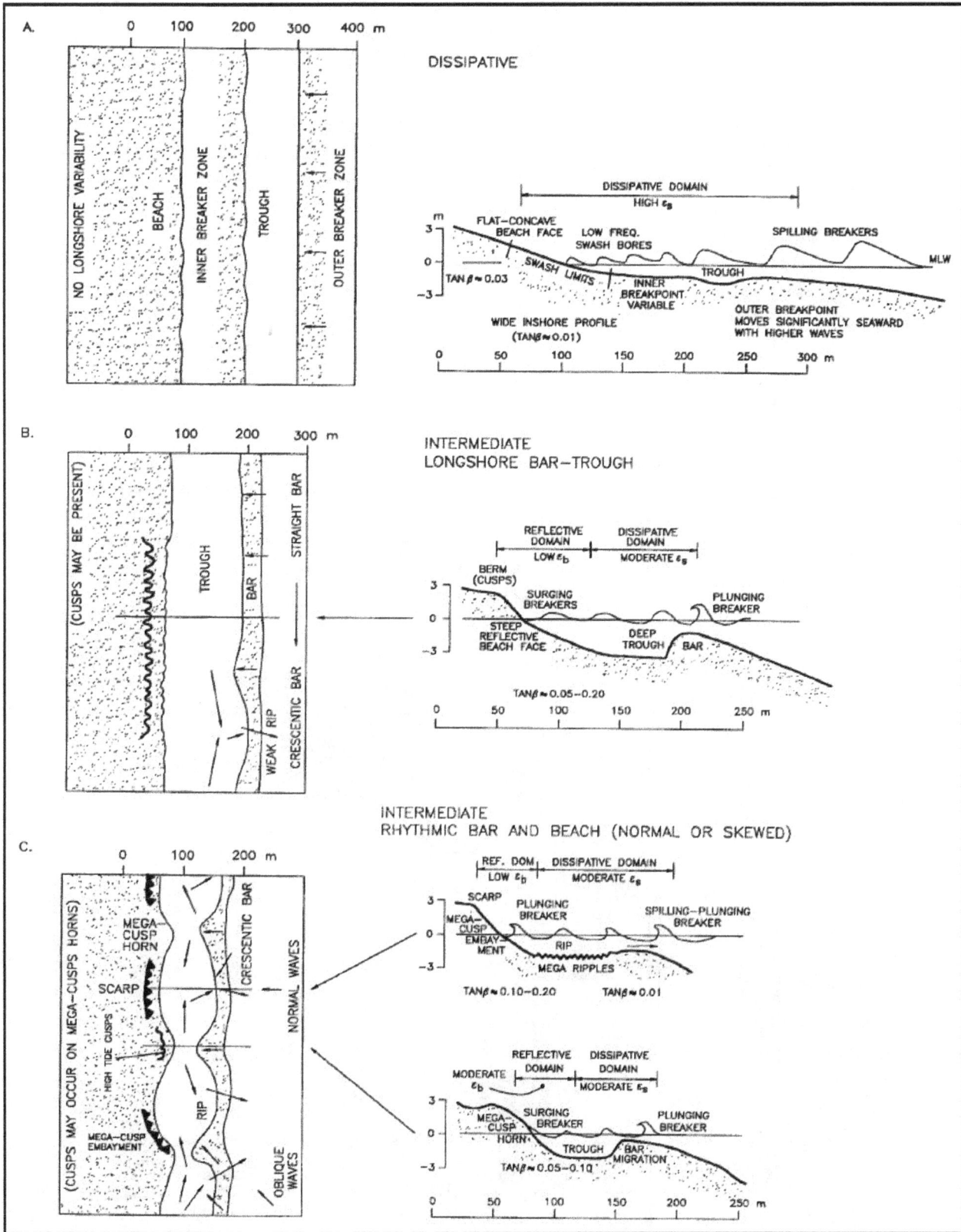

**Figure IV-3-34. Plan and profile views of six major beach stages (adapted from Wright and Short (1984)). Surf-scaling parameter ε is discussed in the text; β represents beach gradient. Dimensions are based on Australian beaches, but morphologic configurations are applicable to other coastlines (Continued)**

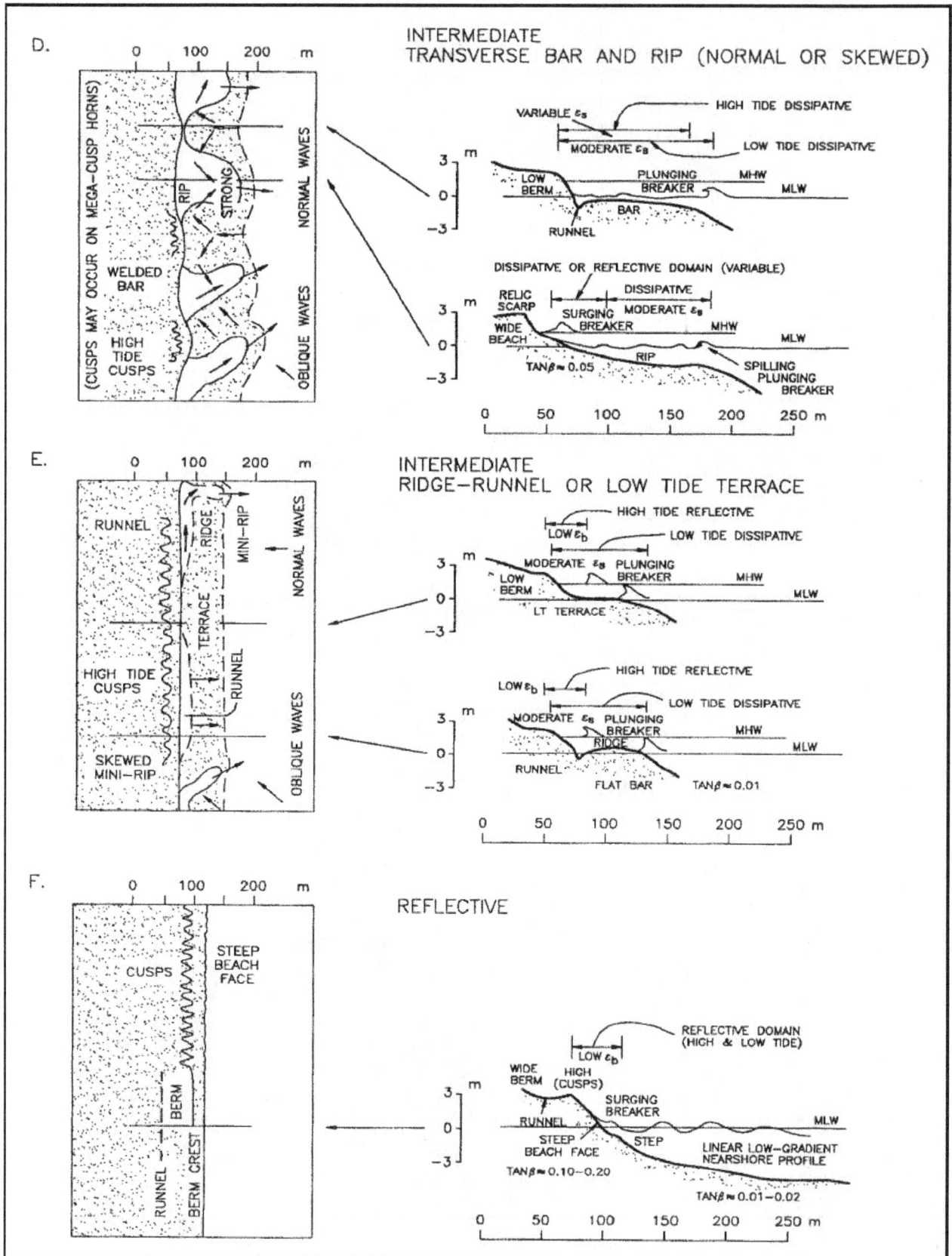

Figure IV-3-34.  (Concluded)

(g) The parameter $\Omega$ depends critically upon $\overline{w_s}$, the sediment fall velocity. It is unclear how the relationships described above apply to shorefaces where the grain size varies widely or where there is a distinct bimodal distribution. For example, many Great Lakes beaches contain material ranging in size from silt and clay to cobble several centimeters in diameter. During storms, not only do wave height and period change, but fine-grain sediment is preferentially removed from the shoreface; therefore, the effective $\overline{w_s}$ may change greatly within a few hours. Further research is needed to understand how Great Lakes beaches change modally and temporally.

(2) Highly dissipative stage (Figure IV-3-34a). The dissipative end of the continuum is analogous to the "storm" or "winter beach" profile described by Bascom (1964) for shores that vary seasonally. The characteristic feature of these beaches is that waves break by spilling and dissipating progressively as they cross a wide surf zone, finally becoming very small at the upper portion of the foreshore (Figure IV-3-35) (Wright and Short 1984). A dissipative surf zone is broad and shallow and may contain two or three sets of bars upon which breakers spill. Longshore beach variability is minimal.

Figure IV-3-35. Example of a dissipative beach: Southern California near San Diego

(3) Highly reflective stage (Figure IV-3-34f). On a fully reflective beach, breakers impinge directly on the shore without breaking on offshore bars (Figures IV-3-36 and 37). As breakers collapse, the wave uprush surges up a steep foreshore. At the bottom of the steep, usually linear beach is a pronounced step composed of coarser material. Seaward of the step, the slope of the bed decreases appreciably. Rhythmic beach cusps are often present in the swash zone. The fully reflective stage is analogous to the fully accreted "summer profile."

(4) Surf-scaling parameter. Morphodynamically, the two end members of the beach state model can be distinguished on the basis of the surf-scaling parameter (Guza and Inman 1975):

**Figure IV-3-36. Example of a reflective sand beach: Newport Beach, California, April 1993**

$$\varepsilon = \frac{a_b \omega^2}{g \tan^2\beta}$$

(IV-3-4)

where

$a_b$ = breaker amplitude

$\omega$ = incident wave radian energy ($2\pi/T$ where $T$ = period)

$g$ = acceleration of gravity

$\beta$ = the gradient of the beach and surf zone

Strong reflection occurs when $\varepsilon \leq 2.0$-$2.5$; this situation defines the highly reflective extreme. When $\varepsilon > 2.5$, waves begin to plunge, dissipating energy. Finally, when $\varepsilon > 20$, spilling breakers occur, the surf zone widens, and turbulent dissipation of wave energy increases with increasing $\varepsilon$.

(5) Intermediate beach stages. These exhibit the most complex morphologies and process signatures.

(a) Longshore bar-trough state (Figure IV-3-34b). This beach form can develop from an antecedent dissipative profile during an accretionary period. Bar-trough relief is higher and the shoreface is much steeper than on the dissipative profile. Initial wave breaking occurs over the bar. However, in contrast to

Figure IV-3-37. Example of a reflective cobble beach: Aldeburgh, Suffolk (facing the North Sea), August 1983. Note the steep berm and the lack of sand-sized sediment. In the background is part of the town of Aldeburgh, which has lost many buildings and churches since the Middle Ages due to erosion

the dissipative beach, the broken waves do not continue to decay after passing over the steep inner face of the bar, but re-form in the deep trough. Low-steepness waves surge up the foreshore; steeper waves collapse or plunge at the base of the foreshore, followed by a violent surge up the subaerial beach (Wright and Short 1984). Runup is relatively high and cusps often occur in the swash zone.

(b) Rhymthic bar and beach (Figure IV-3-34c). Characteristics are similar to the longshore bar-trough state (described above). The distinguishing features of the rhymthic bar and beach state are the regular longshore undulations of the crescentic bar and of the subaerial beach (Figures III-2-23 (ocean) and IV-3-38 (lake)). A weak rip current circulation is often present, with the rips flowing across the narrow portions of the bar. Wright and Short (1984) state that incident waves dominate circulation throughout the surf zone, but subharmonic and infragravity oscillations become important in some regions.

(c) Transverse-bar and rip state (Figure IV-3-34d). This morphology commonly develops in accretionary sequences when the horns of crescentic bars weld to the beach. This results in dissipative transverse bars (sometimes called "mega-cusps") that alternate with reflective, deeper embayments. The dominant dynamic process of this beach state is extremely strong rip circulation, with the seaward-flowing rip currents concentrated in the embayments.

(d) Ridge and runnel/low tide terrace state (Figures IV-3-34e, IV-2-31 (ocean), and IV-3-39 (lake)). This beach state is characterized by a flat accumulation of sand at or just below the low tide level, backed by a steeper foreshore. The beach is typically dissipative at low tide and reflective at high tide.

**Figure IV-3-38. Gravel cusps at St. Joseph, Michigan, November 1993. This is an example of a rhythmic bar and beach on a freshwater coast without tides but subject to irregular seiching**

*e. Processes responsible for shoreface sediment movement.*

(1) Despite intense study for over a century, the subject of sand movement on the shoreface is still poorly understood. Sand is moved by a combination of processes including the following (Pilkey 1993, Wright et al. 1991):

(a) Wave orbital interactions with bottom sediments and with wave-induced longshore currents.

(b) Wind-induced longshore currents.

(c) Turbidity currents.

(d) Rip currents.

(e) Tidal currents.

(f) Storm surge ebb currents.

(g) Gravity-driven currents.

(h) Wind-induced upwelling and downwelling.

(i) Wave-induced upwelling and downwelling.

**Figure IV-3-39. Ridge and runnel north of St. Joseph, Michigan, November 1993. This example shows that these features can be found on lake shores that do not have regular tides**

(j) Gravity-induced downslope transport.

(2) Additional complications are imposed by constantly changing shoreface conditions, as follows:

(a) The relative contributions made by the different transport mechanisms vary over time.

(b) Because of differing regional geological configuration and energy climate, the frequencies of occurrence of the different mechanisms vary with location.

(c) Oscillatory flows normally occur at many frequencies and are superimposed on mean flows and other oscillatory flows of long period.

(3) Middle Atlantic Bight experiments of Wright et al. (1991).

(a) Wright et al. (1991) measured suspended sediment movement, wave heights, and mean current flows at Duck, North Carolina, in 1985 and 1987 and at Sandbridge, Virginia, in 1988 using instrumented tripods. During their study, which included both fair weather and moderate energy conditions, onshore mean flows (interpreted to be related to tides), were dominant over incident waves in generating sediment fluxes. In contrast, during a storm, bottom conditions were strongly dominated by offshore-directed, wind-induced mean flows. Wright et al. attributed this offshore-directed flow to a rise of 0.6 m in mean water level (during this particular storm) and a resultant strong seaward-directed downwelling flow.

(b) Wright et al. (1991) examined the mechanisms responsible for onshore and offshore sediment fluxes across the shoreface. They related two factors explicitly to incoming incident waves:

- Sediment diffusion arising from gradients in wave energy dissipation.

- Sediment advection caused by wave orbital asymmetries.

They found that four other processes may also play important roles in moving sediment:

- Interactions between groupy incident waves and forced long waves.

- Wind-induced upwelling and downwelling currents.

- Wave-current interactions.

- Turbidity currents.

Overall, Wright et al. found that incoming incident waves were of primary importance in bed agitation, while tide- and wind-induced currents were of primary importance in moving sediment. The incoming wave orbital energy was responsible for mobilizing the sand, but the unidirectional currents determined where the sand was going. Surprisingly, cross-shore sediment fluxes generated by mean flows were dominant or equal to sediment fluxes generated by incident waves in all cases and at all times. '

(c) Based on the field measurements, Wright et al. (1991) concluded that "near-bottom mean flows play primary roles in transporting sand across isobaths on the upper shoreface" (p. 49). It is possible that this dominance of mean flows is a feature that distinguished the Middle Atlantic Bight from other shorefaces. The oscillatory (wave) constituents may be proportionately much more important along coasts subject to persistent, high-energy swell, such as the U.S. west coast. Wright et al. also concluded that the directions, rates, and causes of cross-shore sediment flux varied temporally in ways that were only partly predictable with present theory.

   *f.   Sea level change and the Bruun rule.*

(1) General coastal response to changing sea level.[1] Many barrier islands around the United States have accreted vertically during the Holocene rise in global sea level, suggesting that in these areas the supply of sediment was sufficient to allow the beaches to keep pace with the rise of the sea. It is not clear how beaches respond to short-term variations in sea level. Examples of shorter processes include multi-year changes in Great Lakes water levels and multi-month sea level rises associated with the El Niño-Southern Oscillation in the Pacific.

---

[1] Part IV-1 reviewed sea level change and outlined some of the associated coastal effects and management issues. Table IV-1-7 outlined how shoreline advance or retreat at any particular location is a balance between sediment supply and the rate of sea level change. In this section, sea level change is meant in a general sense to be caused by a combination of factors, including eustatic (global) changes and local effects due to vertical movements of the coastal land.

(2) Storm response.

(a) Based on his pioneering research of southern California beaches in the 1940's, Shepard (1950) developed the classic model that there is an onshore-offshore exchange of sediment over winter-summer cycles. Studies since then have shown that this model applies mostly to beaches on swell-dominated coasts where the wave climate changes seasonally (particularly Pacific Ocean coasts) (Carter 1988). Many beaches do *not* show an obvious seasonal cycle. Instead, they erode during storms throughout the year and rebuild during subsequent fair weather periods. Some coasts, like New Jersey, have a seasonal signature, but storms cause such great perturbations that it can take repetitive surveys over many years to extract the seasonal signature.

(b) In some locations, such as the Gulf Coast, infrequent and irregular hurricanes may be the most important dynamic events affecting beaches. Following one of these storms, beach and dune rebuilding may take years (Figure IV-2-10 shows a portion of the Florida/Alabama shore that was damaged by Hurricane Frederic in 1979 and is slowly recovering). Recently, the popular belief that hurricanes are the most important morphodynamic events causing Gulf Coast beach erosion is being reevaluated with the benefit of new field data. Scientists have learned that, cumulatively, winter cold fronts produce significant annual barrier island retreat. Dingler, Reiss, and Plant (1993) monitored Louisiana's Isles Dernieres and found that Hurricane Gilbert (September 1988) produced substantial beach retreat initially, but it actually reduced the average erosion rate by modifying the slope of the shoreface from that produced by cold-front-generated storms. The different responses were related to the scale of the storms. Cold fronts, which individually were small storms, eroded the entire beach to the same degree. Most sand and mud was deposited offshore and only a small percentage of eroded sand was deposited on the backshore because the fronts usually did not raise the sea enough to cause overtopping. Hurricane Gilbert, in contrast, raised sea level substantially such that the primary erosion occurred on the upper beach, and much of the sand was deposited behind the island via overwash processes. Over a 5-year period, the overall effect of this hurricane on the Isles Dernieres was to retard the retreat rate of the island by about 50 percent over that produced by cold fronts alone.

(3) Bruun Rule beach response model.

(a) One of the best-known shoreface response models was proposed by Bruun in 1962 (rederived in Bruun (1988)). Bruun's concept was that beaches adjust to the dominant wave conditions at the site. He reasoned that beaches had to respond in some manner because clearly they had adjusted and evolved historically as sea level had changed. Beaches had not disappeared, they had moved. How was this translation accomplished? Earlier studies of summer/winter beach morphology provided clues that beaches responded even to seasonal changes in wave climate. The basic assumption behind Bruun's model is that with a rise in sea level, the equilibrium profile of the beach and the shallow offshore moves upward and landward. Bruun made several assumptions in his two-dimensional analysis:

- The upper beach erodes because of a landward translation of the profile.

- Sediment eroded from the upper beach is deposited immediately offshore; the eroded and deposited volumes are equal (i.e., longshore transport is not a factor).

- The rise in the seafloor offshore is equal to the rise in sea level. Thus, offshore, the water depth stays constant.

(b) The Bruun Rule can be expressed as (Figure IV-3-40a):

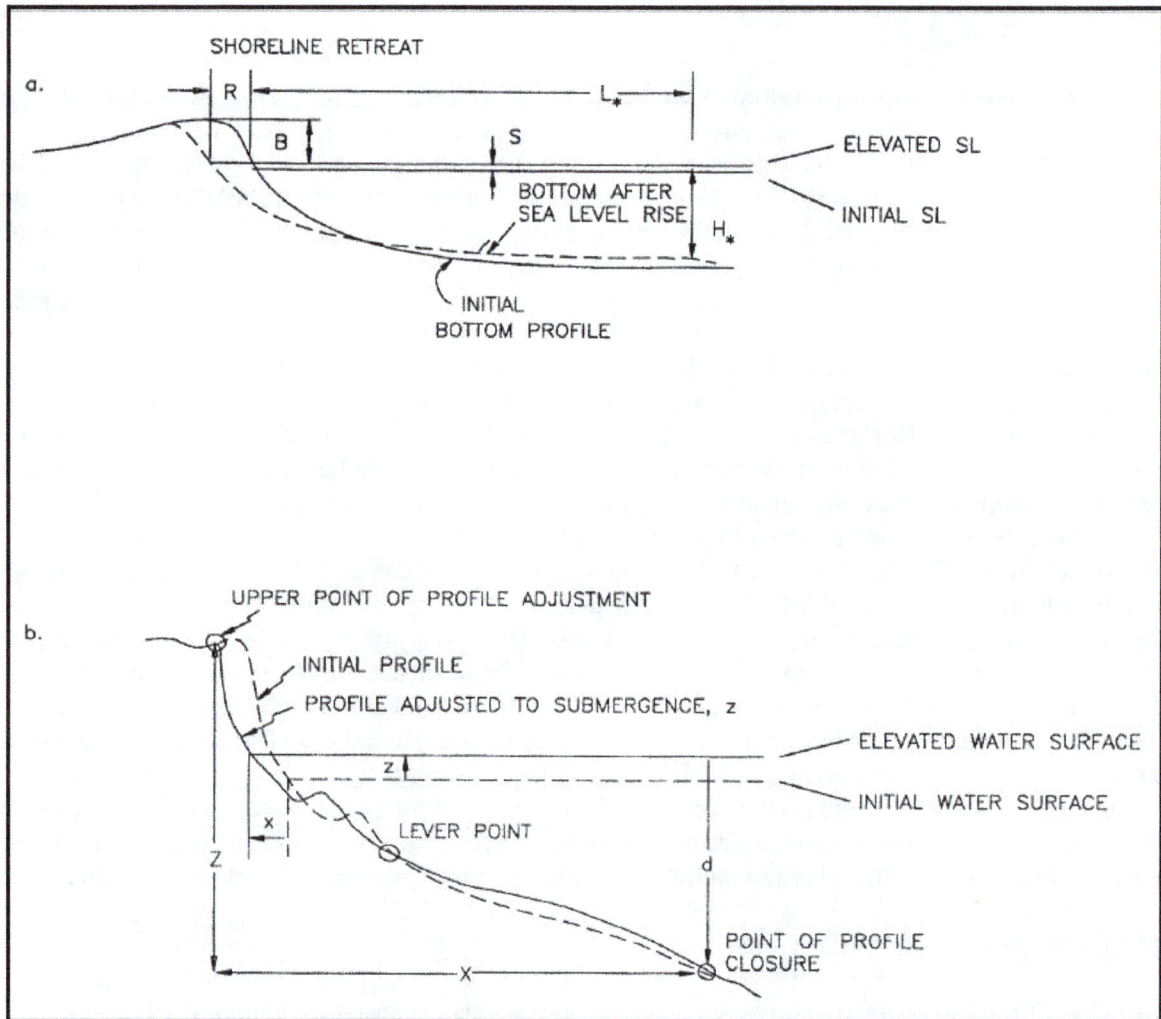

**Figure IV-3-40. (a) Shoreline response to rising sea level (SL) depicted by the Bruun Rule. (b) Simplified nomenclature used by Hands (1983). The sandbar shows that the model is valid for complicated profile shapes**

$$R = \frac{L_*}{B + H_*} S$$

(IV-3-5)

where

$R$ = shoreline retreat

$S$ = increase in sea level

$L_*$ = cross-shore distance to the water depth $H_*$

$B$ = berm height of the eroded area

Hands (1983) restated the Bruun Rule in simplified form:

$$x = \frac{zX}{Z} \qquad\qquad (IV\text{-}3\text{-}6)$$

where $z$ is the change in water level. The ultimate retreat of the profile $x$ can be calculated from the dimensions of the responding profile, $X$ and $Z$, as shown in Figure IV-3-40b. Other expressions for the Bruun Rule are presented in Part III-3-h.

(c) Despite the continued interest in Bruun's concept, there has been only limited use of this method for predictive purposes. Hands (1983) listed several possible reasons for the reluctance to apply this approach:

- Skepticism as to the adequacy of an equilibrium model for explaining short-term dynamic changes.

- Difficulties in measuring sediment lost from the active zone (alongshore, offshore to deep water, and onshore via overwash).

- Problems in establishing a realistic closure depth below which water level changes have no measurable effect on the elevation or slope of the seafloor.

- The perplexity caused by a discontinuity in the profile at the closure depth which appeared in the original and in most subsequent diagrams illustrating the concept.

An additional, and unavoidable, limitation of this sediment budget approach is that it does not address the question of *when* the predicted shore response will occur (Hands 1983). It merely reveals the horizontal distance the shoreline must *ultimately* move to reestablish the equilibrium profile at its new elevation under the assumptions stated in Bruun's Rule.

(d) Hands (1983) demonstrated the geometric validity of the Bruun Rule in a series of figures which show the translation of the profile upward and landward (the figures are two-dimensional; volumes must be based on unit lengths of the shoreline):

- Figure IV-3-41a: The equilibrium profile at the initial water level.

- Figure IV-3-41b: The first translation moves the active profile up an amount $z$ and reestablishes equilibrium depths below the now elevated water level. Hands defines the *active profile* as the zone between the closure depth and the upper point of profile adjustment. The volume of sediment required to maintain the equilibrium water depth is proportional to $X$ (width of the active zone) times $z$ (change in water level).

- Figure IV-3-41c: The required volume of sediment is provided by the second translation, which is a recession (horizontal movement) of the profile by an amount $x$. The amount of sediment is proportional to $x$ times $Z$, where $Z$ is the vertical extent of the active profile from the closure depth to the average elevation of the highest erosion on the backshore.

- Figure IV-3-41d: Equating the volume required by the vertical translation and the volume provided by the horizontal translation yields Equation 3-6. In reality, both translations occur simultaneously, causing the closure point to migrate upslope as the water level rises.

(e) One of the strengths of the Bruun concept is that the equations are valid regardless of the shape of the profile, for example, if bars are present (Figure IV-3-40b). It is important that an offshore distance and depth of closure be chosen that incorporate the entire zone where active sediment transport occurs. Thereby,

a. Equilibrium Profile

Initial Water Surface

Closure Depth

b. Increase of Water Level and Profile Elevations

z

Final
Initial  > Water Surfaces

Required Volume Proportional to Xz

X

c. Recession of Profile

x

Z

Eroded Volume Proportional to xZ

d. Net Results

$$x = \frac{Xz}{Z}$$

Eroded

z

Z

Deposited

x

X

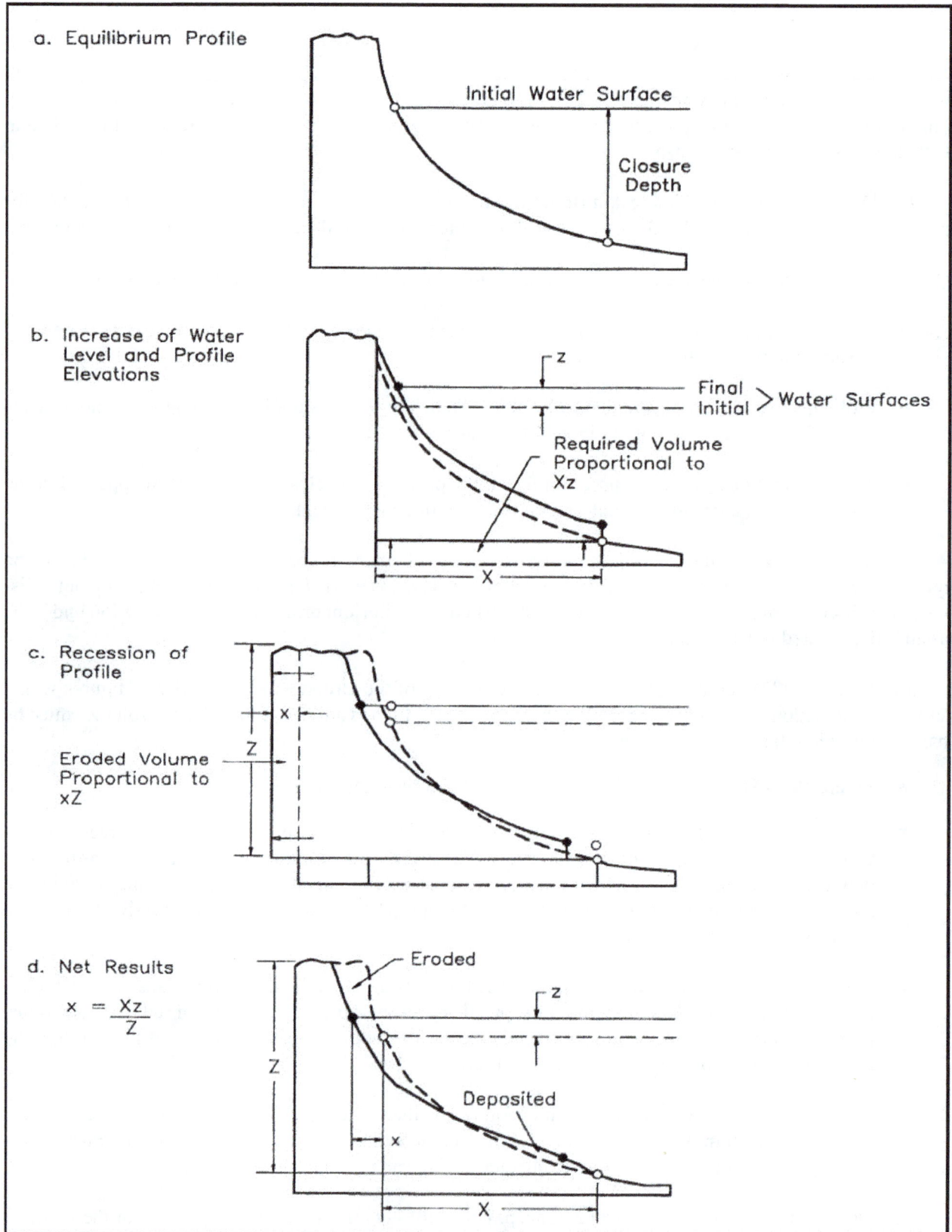

Figure IV-3-41. Profile adjustment in two stages, first vertical, then horizontal, demonstrating the basis for the Bruun Rule (Equation 3-6) (from Hands (1983)). Details discussed in the text

sediment is conserved in spite of the complex processes of local erosion versus deposition as bars migrate (Komar et al. 1991). Another strength is that it is a simple relationship, a geometric conclusion based only on water level. Despite its simplicity and numerous assumptions, it works remarkably well in many settings. Even with its shortcomings, it can be used to predict how beaches can respond to changes in sea level.

(4) Use of models to predict shoreline recession. Although field studies have confirmed the assumptions made by Bruun and others concerning translations of the shoreface, there has been no convincing demonstration that the models can predict shoreline recession rates. Komar et al. (1991) cite several reasons for the inability to use the models as predictive tools:

(a) Existence of a considerable time lag of the beach response following a sustained water level rise (as shown by Hands (1983) for Lake Michigan).

(b) Uncertainty in the selection of the parameters used in the equations (in particular, closure depth).

(c) Local complexities of sediment budget considerations in the sand budget.

(5) Recommendations. More field and laboratory studies are needed to better evaluate the response of beaches to rising (and falling) sea level. For example, it would be valuable to reoccupy the profile lines monitored by Hands (1976, 1979, 1980) in Lake Michigan in the 1970's to determine how the shores have responded to the high water of the mid-1980's and to the subsequent drop in the early 1990's. In addition, conceptual advances need to be incorporated in the theoretical models. How sediment has moved onshore in some locations following sea level rise also needs to be evaluated, because there is evidence that in some areas beach sand compositions reflect offshore rather than onshore sources (Komar et al. 1991).

g. *Equilibrium profiles on sandy coasts.*

(1) General characteristics and assumptions. The existence of an equilibrium shoreface profile (sometimes called equilibrium *beach* profile) is a basic assumption of many conceptual and numerical coastal models. Dean (1990) listed characteristic features of profiles:

(a) Profiles tend to be concave upwards.

(b) Fine sand is associated with mild slopes and coarse sand with steep slopes.

(c) The beach (above the surf zone) is approximately planar.

(d) Steep waves result in milder inshore slopes and a tendency for bar formation.

The main assumption underlying the concept of the shoreface equilibrium profile is that the seafloor is in equilibrium with *average* wave conditions. Presumably, the term *equilibrium* is meant to indicate a situation in which water level, waves, temperature, etc., are held constant for a sufficient time such that the beach profile arrives at a final, stable shape (Larson and Kraus 1989a). Larson (1991) described the profile as: "A beach of specific grain size, if exposed to constant forcing conditions, normally assumed to be short-period breaking waves, will develop a profile shape that displays no net change in time." This concept ignores the fact that, in addition to wave action, many other processes affect sediment transport. These simplifications, however, may represent the real strength of the concept because it has proven to be a useful way to characterize the shape of the shoreface in many locations around the world.

(2) Shape.  Based on studies of beaches in many environments, Bruun (1954) and Dean (1976, 1977) have shown that many ocean beach profiles exhibit a concave shape such that the depth varies as the two-thirds power of distance offshore along the submerged portions:

$$h = Ay^{2/3}$$

(IV-3-7)

where

$h$ = water depth (m) at distance y (m) from the shoreline

$A$ = a scale parameter that depends mainly on sediment characteristics

This surprisingly simple expression asserts, in effect, that beach profile shape can be calculated from sediment characteristics (particle size or fall velocity) alone.  Moore (1982) graphically related the parameter $A$, sometimes called the *profile shape parameter,* to the median grain size $d_{50}$.  Hanson and Kraus (1989) approximated Moore's curve by a series of lines grouped as a function of the median nearshore grain size $d_{50}$ (in mm):

$$A = 0.41(d_{50})^{0.94} \quad , \quad d_{50} < 0.4$$
$$A = 0.23(d_{50})^{0.32} \quad , \quad 0.4 \leq d_{50} < 10.0$$

(IV-3-8)

$$A = 0.23(d_{50})^{0.28} \quad , \quad 10.0 \leq d_{50} < 40.0$$
$$A = 0.46(d_{50})^{0.11} \quad , \quad 40.0 \leq d_{50}$$

Note that A has unit dependence ($m^{1/3}$).  Equation 3-8 is for SI (metric) values only.  Different equations must be used for English units.

Table III-3-3 is a summary of recommended $A$ values, and a more detailed discussion of methods to compute equilibrium profiles is provided in Part III-3.

(3) Discussion of assumptions.  Pilkey et al. (1993), in a detailed examination of the concept of the equilibrium shoreface profile, contended that several assumptions must hold true for the concept to be valid:

(a) *Assumption 1:  All sediment movement is driven by incoming wave orbitals acting on a sandy shoreface.*

This assumption is incorrect because research by Wright et al. (1991) showed that sediment movement on the shoreface is an exceedingly complex phenomenon, driven by a wide range of wave, tidal, and gravity currents.  Even in locations where the wave orbitals are responsible for mobilizing the sand, bottom currents frequently determine where the sand will go.

(b) *Assumption 2:  Existence of closure depth and no net cross-shore (i.e., shore-normal) transport of sediment to and from the shoreface.*

Pilkey et al. (1993) state that this assumption is also invalid because considerable field evidence has shown that large volumes of sand may frequently move beyond the closure depth.  Such movement can occur during both fair weather and storm periods, although offshore-directed storm flows are most likely the prime transport agent. Pilkey et al. cite studies in the Gulf of Mexico that measured offshore bottom currents of up to 200 cm/sec and sediment transport to the edge of the continental shelf. The amount of sediment moved offshore was large, but it was spread over such a large area that the change in sea bed elevation could not be detected by standard profiling methods.[1] Wright, Xu, and Madsen (1994) measured significant across-shelf benthic transport on the inner shelf of the Middle Atlantic Bight during the Halloween storm of 1991.

(c) *Assumption 3: There exists a sand-rich shoreface; the underlying and offshore geology must not play a part in determining the shape of the profile.*

Possibly the most important of the assumptions implicit in the equilibrium profile concept is that the entire profile is sand-rich, without excessive areas of hard bottom or mud within the active profile. Clearly these conditions do not apply in many parts of the world. Coasts that have limited sand supplies, such as much of the U.S. Atlantic margin, are significantly influenced by the geologic framework occurring underneath and in front of the shoreface. Many of the east coast barriers are perched on a platform of ancient sediment. Depending upon the physical state, this underlying platform can act as a subaqueous headland or hardground that dictates the shape of the shoreface profile and controls beach dynamics and the composition of the sediment. Niederoda, Swift, and Hopkins (1985) believed that the seaward-thinning and fining veneer of modern shoreface sediments is ephemeral and is easily removed from the shoreface during major storms. During storms, Holocene and Pleistocene strata cropping out on the shoreface provide the immediate source of the bulk of barrier sands. Swift (1976) used the term *shoreface bypassing* to describe the process of older units supplying sediment to the shoreface of barrier islands. Pilkey et al. (1993) contend that:

...a detailed survey of the world's shorefaces would show that the sand rich shoreface required by the equilibrium profile model is an exception rather than the rule. Instead, most shorefaces are underlain by older, consolidated or semi-consolidated units covered by only a relatively thin veneer of modern shoreface sands. These older units are a primary control on the shape of the shoreface profile. The profile shape is not determined by simple wave interaction with the relatively thin sand cover. Rather, the shape of the shoreface in these sediment poor areas is determined by a complex interaction between underlying geology, modern sand cover, and highly variable (and often highly diffracted and refracted) incoming wave climate. (p. 271)

(d) *Assumption 4: If a shoreface is, in fact, sand-rich, the smoothed profile described by the equilibrium profile equation (ignoring bars and troughs) must provide a useful approximation of the real shoreface shape.*

In addressing this assumption, Pilkey et al. (1993) cited studies conducted on the Gold Coast, in Queensland, Australia. The Gold Coast shoreface is sand-rich to well beyond a depth of 30 m. Without being directly influenced by underlying geology, the shoreface is highly dynamic. As a consequence, the Gold Coast shoreface shape cannot be described by one equilibrium profile; rather, it is best described by an ever-changing regime profile. Pilkey et al. concluded:

The local shoreface profile shapes are entirely controlled by relative wave energy "thresholds"; for the sediment properties have not changed at all. Thus principal changes to the shoreface profiles of the Gold Coast are driven by wave power history with some modification by currents, and not by sediment size, or its parameter *A*, as defined within the equilibrium profile concept. (p. 272).

---

[1] This latter statement underscores how important it is to develop improved methods to detect and measure sediment movement in deep water.

(4) General comments.

(a) The idea of a profile only adjusting to waves is fundamentally wrong as shown by Wright et al. (1991) and others. However, although the physical basis for the equilibrium profile concept is weak, critics of this approach have not proven that it always results in highly erroneous answers.

(b) Before the use of the equilibrium profile, coastal engineers had no way to predict beach change other than using crude approximations (e.g., sand loss of 1 cu yd/ft of beach retreat). The approximations were inadequate. Surveys from around the world have shown that shoreface profiles display a characteristic shape that differs with locality but is relatively stable for a particular place (i.e., Duck, North Carolina). With many caveats (which are usually stated, then ignored), a profile can be reasonably represented by the equilibrium equation. The fit between the profile and the real seafloor on a daily, seasonal, and storm variation basis may not be perfect, but the differences may not matter in the long term.

(c) One critical problem for coastal engineers is to predict what a sequence of waves (storm) will do to a locality when little is known about the particular shape of the pre-storm beach. For this reason, numerical models like SBEACH (Larson and Kraus 1989a), despite their reliance on the equilibrium profile concept, are still useful. The models allow a researcher to explore storm impact on a location using a general approximation of the beach. The method is very crude - however, the resulting numbers are of the right order of magnitude when compared with field data from many locations.

(d) Answers from the present models are not exact, and researchers still have much to learn about the weakness of the models and about physical processes responsible for the changes. Nevertheless, the models do work and they do provide numbers that are of the correct magnitudes when run by careful operators. Users of shoreface models must be aware of the limitations of the models and of special conditions that may exist at their project sites. In particular, profile-based numerical models are likely to be inadequate in locations where processes other than wave-orbital transport predominate.

*h . Depth of closure.*

(1) Background.

(a) *Depth of closure* is a concept that is often misinterpreted and misused. For engineering practice, depth of closure is commonly defined as the minimum water depth at which no measurable or significant change in bottom depth occurs (Stauble et al. 1993). The word *significant* in this definition is important because it leaves considerable room for interpretation. "Closure" has erroneously been interpreted to mean the depth at which no sediment moves on- or offshore, although numerous field studies have verified that much sediment moves in deep water (Wright et al. 1991). Another complication is introduced by the fact that it is impossible to define a single depth of closure for a project site because "closure" moves depending on waves and other hydrodynamic forces. Therefore, it is invalid to assume that "closure" is a single fixed depth at a project site or a stretch of coastline.

(b) Closure depth is used in a number of applications such as the placement of mounds of dredged material, beach fill, placement of ocean outfalls, and the calculation of sediment budgets.

(2) Energy factors. As discussed above, the primary assumption behind the concept of the shoreface equilibrium profile is that sediment movement and the resultant changes in bottom elevation are a function of wave properties and sediment grain size. Therefore, the active portion of the shoreface varies in width throughout the year depending on wave conditions. In effect, "closure" is a time-dependent quantity that may be predicted based on wave climatology or may be interpreted statistically using profile surveys.

(3) Time considerations. The energy-dependent nature of the active portion of the shoreface requires us to consider return period. The closure depth that accommodates the 100-year storm will be much deeper than one that merely needs to include the 10-year storm. Therefore, a closure depth must be chosen in light of a project's engineering requirements and design life. For example, if a berm is to be built in deep water where it will be immune from wave resuspension, what is the minimum depth at which it should be placed? This is an important question because of the high costs of transporting material and disposing of it at sea. It would be tempting to use a safe criterion such as the 100- or 500-year storm, but excessive costs may force the project engineer to consider a shallower site that may be stable only for shorter return period events.

(4) Predictive methods.

(a) Hallermeier (1977, 1978, 1981a, 1981b, 1981c), using laboratory tests and limited field data, introduced equations to predict the limits of extreme wave-related sediment movement. He calculated two limits, $d_\ell$ and $d_i$, that included a buffer region on the shoreface called the shoal zone. Landward of $d_\ell$, significant alongshore transport and intense onshore-offshore sediment transport occur (the littoral zone). Within the shoal zone, expected waves have neither a strong nor a negligible effect on the sandy bed during a typical annual cycle of wave action. Seaward of $d_i$, only insignificant onshore-offshore transport by waves occurs. The deeper limit was based on the median nearshore storm wave height (and the associated wave period). The boundary between the shoal zone and the littoral zone ($d_\ell$) as defined represents the annual depth of closure. Hallermeier (1978) suggested an analytical approximation, using linear wave theory for shoaling waves, to predict an *annual* value of $d_\ell$ :

$$d_\ell = 2.28 H_e - 68.5 \left(\frac{H_e^2}{g T_e^2}\right) \qquad \text{(IV-3-9)}$$

where

$d_\ell$ = annual depth of closure below mean low water

$H_e$ = non-breaking significant wave height that is exceeded 12 hr per year (0.137 % of the time)

$T_e$ = associated wave period

$g$ = acceleration due to gravity

According to Equation 3-9, $d_\ell$ is primarily dependent on wave height with an adjustment for wave steepness. Hallermeier (1978) proposed using the 12-hr exceeded wave height, which allowed sufficient duration for "moderate adjustment towards profile equilibrium." Equation 3-9 is based on quartz sand with a submerged density of $\gamma' = 1.6$ and a median diameter between 0.16 and 0.42 mm, which typifies conditions in the nearshore for many beaches. If the grain size is larger than 0.42 mm, Equation 3-9 may not be appropriate. Because $d_\ell$ was derived from linear wave theory for shoaling waves, $d_\ell$ must be seaward of the influence of intense wave-induced nearshore circulation. However, because of various factors, Hallermeier (1978) "proposed that the calculated $d_\ell$ be used as a minimum estimate of profile close-out depth with respect to low(er) tide level." Because tidal or wind-induced currents may increase wave-induced near-bed flow velocities, Hallermeier suggested using mean low water (mlw) as a reference water level to obtain a conservative depth of closure. Note that Hallermeier's equations critically depend on the quality of wave data at a site. The reader is cautioned that Hallermeier's equations can be expressed in various forms depending on the assumptions made, the datums used as reference levels, and available wave data. The reader is referred to his original papers for clarification and for details of his assumptions. The equations may not be applicable at sites where currents are more important at moving sand than wave-induced flows.

(b) At the Lake Michigan sites that Hands (1983) surveyed, the closure depth was equal to about twice the height of the 5-year return period wave height ($H_5$):

$$Z \approx 2H_5 \tag{IV-3-10}$$

In the absence of strong empirical evidence as to the correct closure depth, this relationship is recommended as a rule of thumb to estimate the 5-year profile response under Great Lakes conditions. The return period of the wave height should approximate the design life of interest. For example, the 20-year closure depth would be estimated by doubling the 20-year return period wave height ($Z \approx 2H_{20}$).

(5) Empirical determination.

(a) When cross-shore surveys covering several years are available for a project site, closure is best determined by plotting and analyzing the profiles. The closure depth computed in this manner reflects the influence of storms as well as of calmer conditions. Kraus and Harikai (1983) evaluated the depth of closure as the minimum depth where the standard deviation in depth change decreased markedly to a near-constant value. Using this procedure, they interpreted the landward region where the standard deviation increased to be the active profile where the seafloor was influenced by gravity waves and storm-driven water level changes. The offshore region of smaller and nearly constant standard deviation was primarily influenced by lower frequency sediment-transporting processes such as shelf and oceanic currents (Stauble et al. 1993). It must be noted that the smaller standard deviation values fall within the limit of measurement accuracy. This suggests that it is not possible to specify a closure depth unambiguously because of operational limits of offshore profiling hardware and procedures.

(b) An example of how closure was determined empirically at Ocean City, Maryland, is shown in Figure IV-3-42 (from Stauble et al. (1993)). A clear reduction in standard deviation occurs at a depth of about 5.5 - 6 m. Above the ~5.5 m depth, the profile exhibits large variability, indicating active wave erosion, deposition, and littoral transport. Deeper (and seaward) of this zone, the lower and relatively constant deviation of about 7 - 10 cm is within the measurement error of the sled surveys. Nevertheless, despite the inability to precisely measure seafloor changes in this offshore region, it is apparent that less energetic erosion and sedimentation take place here than in water shallower than ~5.5 m. This does not mean that there is no sediment transport in deep water, just that the sled surveys are unable to measure it. For the 5.6 km of shore surveyed at Ocean City, the depth of closure ranged between 5.5 and 7.5 m. Scatter plots indicated that the average closure depth was 6 m.

(c) Presumably, conducting surveys over a longer time span at Ocean City would reveal seafloor changes deeper than ~6 m, depending on storms that passed the region. However, Stauble et al. (1993) noted that the "Halloween Storm" of October 29 to November 2, 1991 generated waves of peak period ($T_p$) 19.7 sec, extraordinarily long compared to normal conditions along the central Atlantic coast. Therefore, the profiles may already reflect the effects of an unusually severe storm.

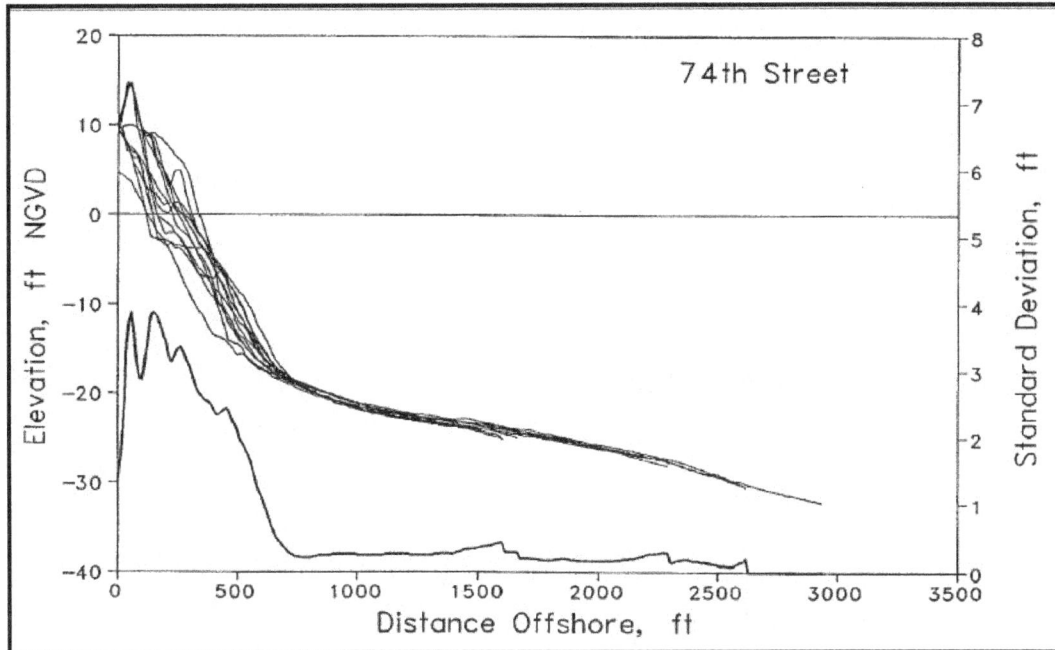

**Figure IV-3-42. Profile surveys and standard deviation of seafloor elevation at 74th Street, Ocean City, Maryland (from Stauble et al. (1993)). Surveys conducted from 1988 to 1992. Large changes above the datum were caused by beach fill placement and storm erosion. Figure discussed in the text**

(d) Figure IV-3-43 is an example of profiles from St. Joseph, Michigan, on the east shore of Lake Michigan. Along Line 14, dramatic bar movement occurs as far as 760 m offshore to a depth of 7.6 m with respect to International Great Lakes Datum (IGLD) 1985. This is where an abrupt decrease in standard deviation of lake floor elevation occurs and can be interpreted as closure depth. In September 1992, the mean water surface was 0.51 m above IGLD 85. Therefore, closure was around 7.9 - 8.2 m below *water* level.

(e) In the Great Lakes, water levels fluctuate over multi-year cycles. This raises some fundamental difficulties in calculating closure based on profile surveys. Presumably, during a period of high lake level, the zone of active sand movement would be higher on the shoreface than during a time of low lake level (this assumes similar wave conditions). Therefore, the depth where superimposed profiles converge should reflect the *deepest* limit of active shoreface sand movement. This would be a conservative value, but *only with respect to the hydrologic conditions that occurred during the survey program.* Presumably, if lake level dropped further at a later date, sediment movement might occur deeper on the shoreface. This suggests that closure on the lakes should be chosen to reflect the *lowest* likely water level that is expected to occur during the life of a project. (Note that this consideration does not arise on ocean coasts because year-to-year changes in relative sea level are minor, well within the error bounds of sled surveys. Sea level does change throughout the year because of thermal expansion, freshwater runoff, and other factors as discussed in Part IV-1, but the multi-year mean is essentially stable.) In summary, determining closure depth in the Great Lakes is problematic because of changing water levels, and more research is needed to develop procedures that accommodate these non-periodic lake level fluctuations.

(f) The variation of closure depth at approximately 100 profile lines along the south shore of Long Island is plotted in Figure IV-3-44. Generally the depth increases towards the east, with Rockaway Beach averaging 5.0 m below NGVD and the Montauk zone averaging 7.6 m. These values were based on

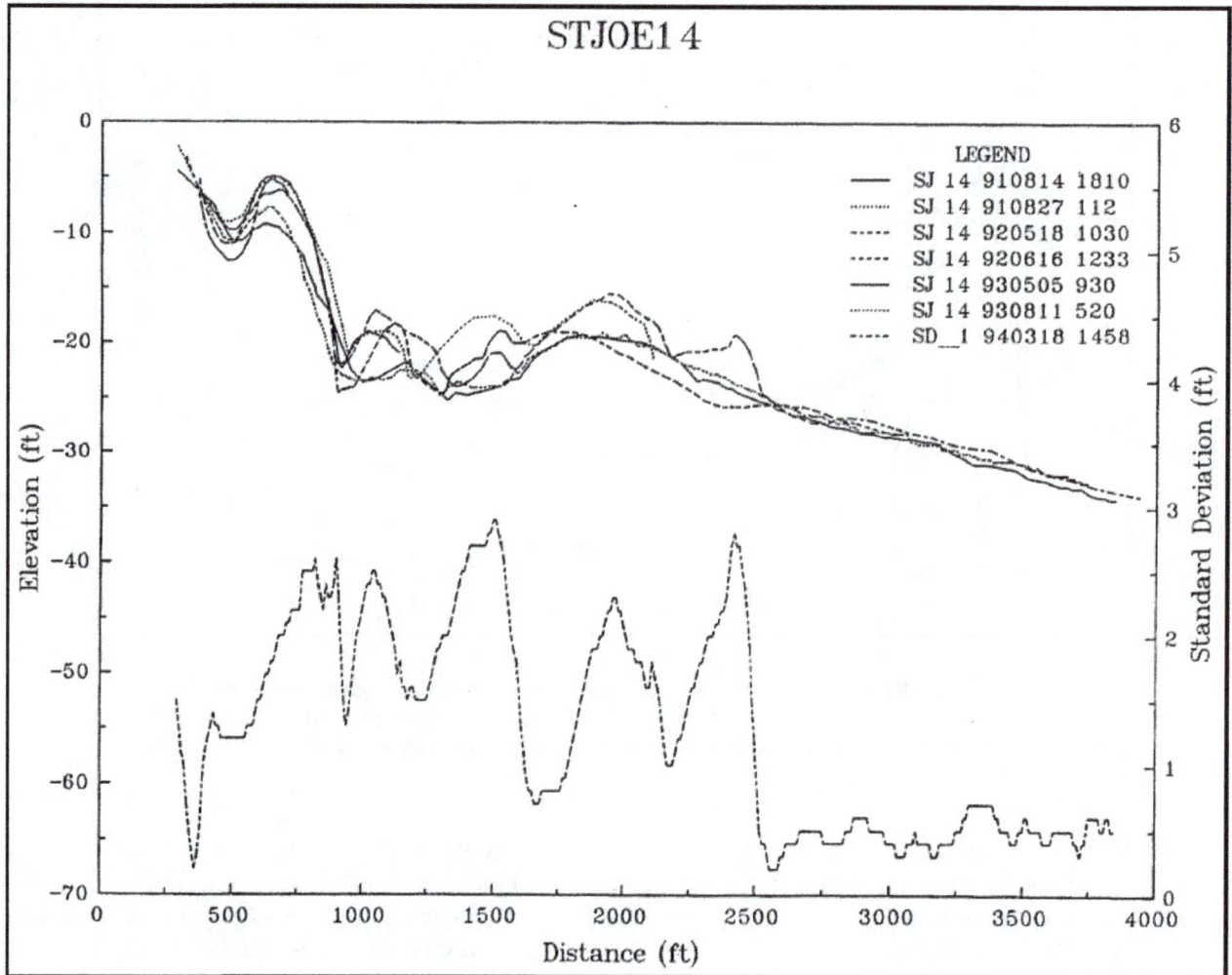

**Figure IV-3-43. Profile surveys and standard deviation of lake floor elevation at St. Joseph, Michigan, on the east shore of Lake Michigan. Profiles are referenced to International Great Lakes Datum (IGLD) 1985. Surveys conducted between 1991 and 1994 (Nairn et al. 1997). Figure discussed in the text**

surveys in 1995 and 1996 (Morang 1998). The depth increase toward the east is expected because wave energy is greater there than at the west end of Long Island.

*i. Longshore sediment movement.*

The reader is referred to Part III-2 and to *Coastal Sediment Transport* (EM 1110-2-1502) for a detailed treatment of longshore transport.

*j. Summary.*

(1) A model of shoreface morphodynamics for micro- and low-mesotidal sandy coasts has been developed by Wright and Short (1984). The six stages of the model (Figure IV-3-34) illustrate the response of sandy beaches to various wave conditions.

(2) Sediment movement on the shoreface is a very complicated phenomenon. It is a result of numerous hydrodynamic processes, among which are: (a) wave orbital interactions with bottom sediments and with

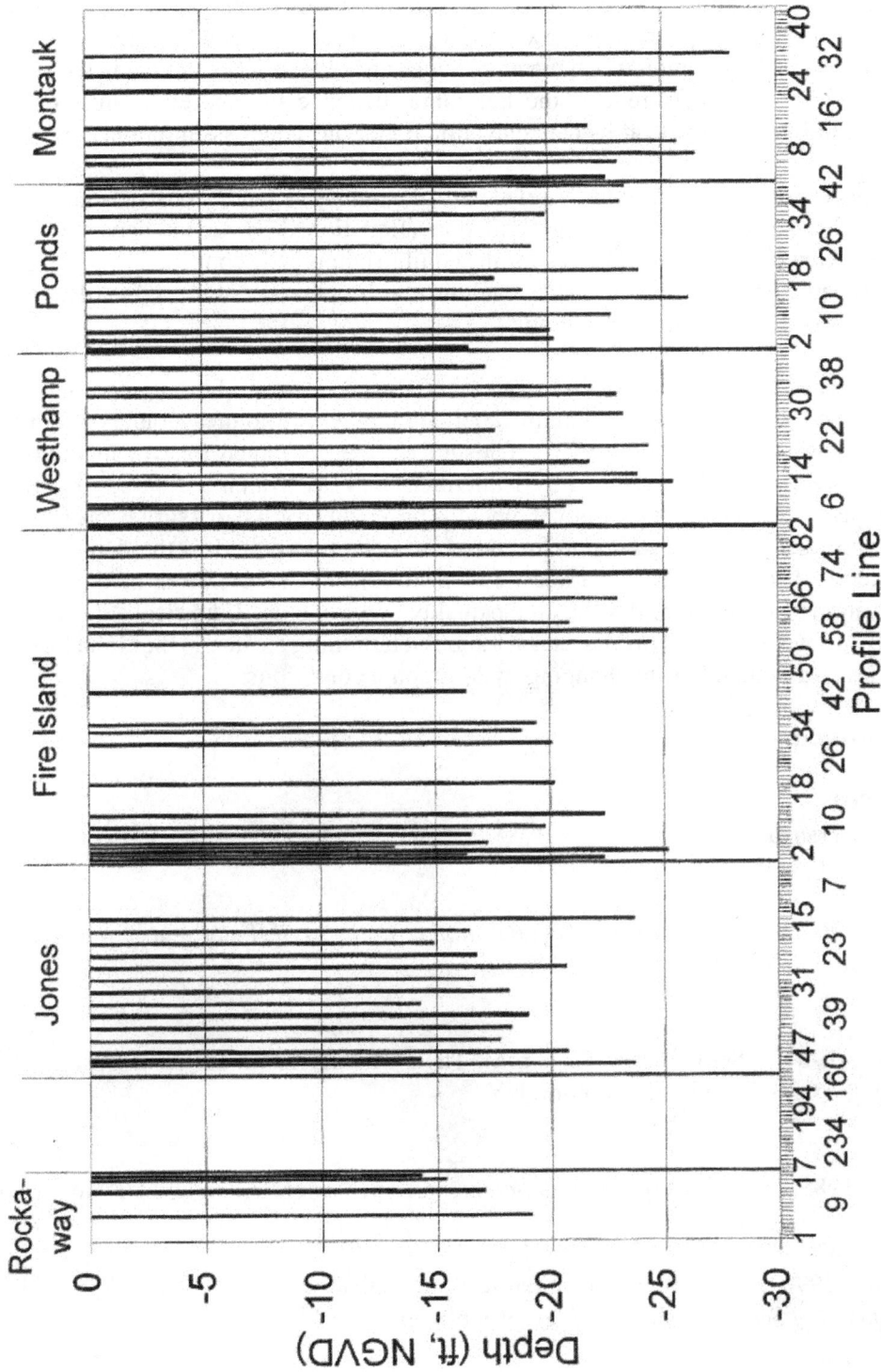

Figure IV-3-44. Variation in short-term closure depth along the south shore of Long Island, New York, computed for four survey dates in 1995 and 1996

wave-induced longshore currents; (b) wind-induced longshore currents; (c) rip currents; (d) tidal currents; (e) storm surge ebb currents; (f) gravity-driven currents; (g) wind-induced upwelling and downwelling; (h) wave-induced upwelling and downwelling; and (i) gravity-induced downslope transport.

(3) The Bruun Rule (Equation 3-5 or 3-6) is a model of shoreface response to rising sea level. Despite the model's simplicity, it helps explain how barriers have accommodated rising sea level by translating upward and landward. A limitation is that the model does not address *when* the predicted shore response will occur (Hands 1983). It merely reveals the horizontal distance the shoreline must *ultimately* move to reestablish the equilibrium profile at its new elevation under the stated assumptions.

(4) The concept of the equilibrium shoreface profile applies to sandy coasts primarily shaped by wave action. It can be expressed by a simple equation (Equation 3-7) which depends only on sediment characteristics. Although the physical basis for the equilibrium profile concept is weak, it is a powerful tool because models based on the concept produce resulting numbers that are of the right order of magnitude when compared with field data from many locations.

(5) *Closure* is a concept that is often misinterpreted and misused. For engineering practice, depth of closure is commonly defined as the minimum water depth at which no measurable or significant change in bottom depth occurs (Stauble et al. 1993). Closure can be computed by two methods: (a) analytical approximations such as those developed by Hallermeier (1978), which are based on wave statistics at a project site (Equation 3-10); or (b) empirical methods based on cross-shore survey profile data. When profiles are superimposed, a minimum value for closure can be interpreted as the depth where the standard deviation in depth change decreases markedly to a near-constant value. Both methods have weaknesses. Hallermeier's analytical equations depend on the quality of wave data. Empirical determinations depend on the availability of several years of profile data at a site. Determining closure in the Great Lakes is problematic because lake levels fluctuate due to changing hydrographic conditions.

## IV-3-6. References

**EM 1110-2-1502**
Coastal Littoral Transport

**Allen 1968**
Allen, J. R. L. 1968. *Current Ripples: Their Relation to Patterns of Water and sediment Movement,* North Holland, Amsterdam, Netherlands.

**Allen 1984**
Allen, J. R. L. 1984. "Sedimentary Structures, Their Character and Physical Basis," *Developments in Sedimentology*, Vol 30, Elsevier, New York, NY.

**Allen 1985**
Allen, J. R. L. 1985. *Principles of Physical Sedimentology,* George Allen and Unwin, London, UK.

**Ashley 1990**
Ashley, G. M. 1990. "Classification of Large-Scale Subaqueous Bedforms: A New Look at an Old Problem," *Journal of Sedimentary Petrology*, Vol 60, pp 363-396.

**Aubrey and Speer 1984**
Aubrey, D. G., and Speer, P. E. 1984. "Updrift Migration of Tidal Inlets," *Journal of Geology*, Vol 92, pp 531-546.

**Aubrey and Weishar 1988**
Aubrey, D. G., and Weishar, L., eds. 1988. *Hydrodynamics and Sediment and Dynamics of Tidal Inlets*, Lecture Notes on Coastal and Estuarine Studies, Vol 29, Springer-Verlag, New York, NY.

**Baeteman 1994**
Baeteman, C. 1994. Subsidence in coastal lowlands due to groundwater withdrawal: The geological approach, "Coastal Hazards, Perception, Susceptibility and Mitigation," C. W., Finkl, Jr., ed., *Journal of Coastal Research* Special Issue No. 12, pp 61-75.

**Barwis 1976**
Barwis, J. H. 1976. "Annotated Bibliography on the Geologic, Hydraulic, and Engineering Aspects of Tidal Inlets," General Investigation of Tidal Inlets Report 4, U.S. Army Engineer Waterways Experiment Station, Vicksburg, MS.

**Bascom 1964**
Bascom, W. 1964. *Waves and Beaches, the Dynamics of the Ocean Surface*, Doubleday & Co., Garden City, NY.

**Bass, Fulford, Underwood, and Parson 1994**
Bass, G. P., Fulford, E. T., Underwood, S. G., and Parson, L. E. 1994. "Rehabilitation of the South Jetty, Ocean City, Maryland," Technical Report CERC 94-6, U.S. Army Engineer Waterways Experiment Station, Vicksburg, MS.

**Boothroyd 1985**
Boothroyd, J. C. 1985. "Tidal Inlets and Tidal Deltas," *Coastal Sedimentary Environments*, 2nd ed., R. A. Davis, ed., Springer-Verlag, New York, NY, pp 445-532.

**Bruno, Yavary, and Herrington 1998**
Bruno, M. S., Yavary, M., and Herrington, T. O. 1998. "The influence of a stabilized inlet on adjacent shorelines: Manasquan, New Jersey," *Shore & Beach,* Vol. 66, No. 2, pp 19-25.

**Bruun 1954**
Bruun, P. 1954. "Coast Erosion and the Development of Beach Profiles," Technical Memorandum No. 44, Beach Erosion Board, U.S. Army Engineer Waterways Experiment Station, Vicksburg, MS.

**Bruun 1966**
Bruun, P. 1966. *Tidal Inlets and Littoral Drift*, Universitets-forlaget, Norway (in English).

**Bruun 1988**
Bruun, P. 1988. "The Bruun Rule of Erosion by Sea-Level Rise: A Discussion of Large-Scale Two- and Three-Dimensional Usages," *Journal of Coastal Research*, Vol 4, pp 627-648.

**Bruun and Gerritsen 1959**
Bruun, P., and Gerritsen, F. 1959. "Natural By-passing of Sand at Coastal Inlets," *Journal of Waterways and Harbors Division,* American Society of Civil Engineers, New York, pp 75-107.

**Bruun and Gerritsen 1961**
Bruun, P., and Gerritsen, F. 1961. Stability of Coastal Inlets," *Proceedings of the Seventh Conference on Coastal Engineering,* August 1960, The Hague, Netherlands, J. W. Johnson, ed., Council on Wave Research, University of California, Berkeley, CA, pp 386-417.

**Carter 1988**
Carter, R. W. G. 1988. *Coastal Environments: An Introduction to the Physical, Ecological, and Cultural Systems of Coastlines,* Academic Press, London, UK.

**Castañer 1971**
Castañer, P. F. 1971. "Selected Bibliography on the Engineering Characteristics of Coastal Inlets," Report HEL 24-7, Hydraulic Engineering Laboratory, University of California, Berkeley, CA.

**Chapman 1974**
Chapman, V. J. 1974. "Salt Marshes and Salt Deserts of the World," *Ecology of Halophytes*, R. J. Reimold and W. H. Queen, eds., Academic Press, New York, pp 3-19.

**Chasten 1992**
Chasten, M. A. 1992. "Coastal Response to a Dual Jetty System at Little River Inlet, North and South Carolina," Miscellaneous Paper CERC-92-2, U.S. Army Engineer Waterways Experiment Station, Vicksburg, MS.

**Chasten and Seabergh 1992**
Chasten, M. A., and Seabergh, W. C. 1992. "Engineering Assessment of Hydrodynamics and Jetty Scour at Little River Inlet, North and South Carolina," Miscellaneous Paper CERC-92-10, U.S. Army Engineer Waterways Experiment Station, Vicksburg, MS.

**Coastal Barriers Study Group 1988**
Coastal Barriers Study Group. 1988. Report to Congress: Coastal Barrier Resources System, Recommendations for Additions to or Deletions from the Coastal Barrier Resources System, (in 22 Volumes covering the United States and territories), U.S. Department of the Interior, Washington, DC.

**Coleman 1988**
Coleman, J. M. 1988. "Dynamic Changes and Processes in the Mississippi River Delta," *Bulletin of the Geological Society of America,* Vol 100, pp 999-1015.

**Coleman and Garrison 1977**
Coleman, J. M., and Garrison, L. E. 1977. "Geological Aspects of Marine Slope Instability, Northwestern Gulf of Mexico," *Marine Geotechnology,* Vol 2, pp 9-44.

**Coleman and Wright 1975**
Coleman, J. M., and Wright, L. D. 1975. "Modern River Deltas: Variability of Process and Sand Bodies," *Deltas, Models for Exploration,* M. L. Broussard, ed., Houston Geological Society, Houston, TX, pp 99-149.

**Cronin 1975**
Cronin, L. E., ed. 1975. *Estuarine Research*, Academic Press, New York (2 volumes).

**Davis 1985**
Davis, R. A., Jr., ed. 1985. *Coastal Sedimentary Environments,* 2nd ed., Springer-Verlag, New York.

**Davis and Ethington 1976**
Davis, R. A., Jr., and Ethington, R. L. 1976. *Beach and Nearshore Sedimentation*, Society of Economic Paleontologists and Mineralogists (SEPM) Special Publication No. 24, Tulsa, OK.

**Dean 1976**
Dean, R. G. 1976. "Beach Erosion: Causes, Processes, and Remedial Measures," *CRC Reviews in Environmental Control*, CRC Press Inc., Boca Raton, FL, Vol 6, Issue 3, pp 259-296.

**Dean 1977**
Dean, R. G. 1977. "Equilibrium Beach Profiles - U.S. Atlantic and Gulf Coasts," Ocean Engineering Report No. 12, University of Delaware, Newark, pp 1-45.

**Dean 1987**
Dean, R. G. 1987. "Coastal Sediment Processes: Toward Engineering Solutions," *Proceedings of Coastal Sediments '87*, American Society of Civil Engineers, New York, pp 1-24.

**Dean 1988**
Dean, R. G. 1988. "Sediment Interaction at Modified Coastal Inlets: Processes and Policies," *Hydrodynamics and Sediment Dynamics of Tidal Inlets*, Lecture Notes on Coastal and Estuarine Studies, D. G. Aubrey and L. Weishar, eds., Vol 29, Springer-Verlag, New York, pp 412-439.

**Dean 1990**
Dean, R. G. 1990. "Equilibrium Beach Profiles: Characteristics and Applications," Report UFL/COEL-90/001, Coastal & Oceanographic Engineering Department, University of Florida, Gainesville, FL.

**Dingler, Reiss, and Plant 1993**
Dingler, J. R., Reiss, T. E., and Plant, N. G. 1993. "Erosional Patterns of the Isles Dernieres, Louisiana, in Relation to Meteorological Influences," *Journal of Coastal Research*, Vol 9, No. 1, pp 112-125.

**Douglass 1987**
Douglass, S. L. 1987. "Coastal Response to Navigation Structures at Murrells Inlet, South Carolina," Technical Report CERC-87-2, U.S. Army Engineer Waterways Experiment Station, Vicksburg, MS.

**Escoffier 1940**
Escoffier, F. F. 1940. "The Stability of Tidal Inlets," *Shore and Beach*, Vol 8, pp 114-115.

**Escoffier 1977**
Escoffier, F. F. 1977. "Hydraulics and Stability of Tidal Inlets," General Investigation of Tidal Inlets Report 13, U.S. Army Engineer Waterways Experiment Station, Vicksburg, MS.

**FitzGerald 1988**
FitzGerald, D. M. 1988. "Shoreline Erosional-Depositional Processes Associated with Tidal Inlets," *Hydrodynamics and Sediment Dynamics of Tidal Inlets*, Lecture Notes on Coastal and Estuarine Studies, D. G. Aubrey and L. Weishar, eds., Vol 29, Springer-Verlag, New York, pp 186-225.

**FitzGerald and Nummedal 1983**
FitzGerald, D. M., and Nummedal, D. 1983. "Response Characteristics of an Ebb-Dominated Tidal Inlet Channel," *Journal of Sedimentary Petrology*, Vol 53, No. 3, pp 833-845.

**FitzGerald, Hubbard, and Nummedal  1978**
FitzGerald, D. M., Hubbard, D. K., and Nummedal, D.  1978.  "Shoreline Changes Associated with Tidal Inlets along the South Carolina Coast," *Proceedings Coastal Zone '78*, American Society of Civil Engineers, New York, pp 1973-1994.

**FitzGerald, Penland, and Nummedal  1984**
FitzGerald, D. M., Penland, S., and Nummedal, D.  1984.  "Control of Barrier Island Shape by Inlet Sediment Bypassing:  Ease Friesian Islands, West Germany," *Marine Geology*, Vol 60, pp 355-376.

**Gorman, Morang, and Larson 1998**
Gorman, L., Morang, A., and Larson, R. 1998.  "Monitoring the Coastal Environment; Part IV: Mapping, Shoreline Changes, and Bathymetric Analysis," *Journal of Coastal Research* 14(1), 61-92.

**Graf  1984**
Graf, W. H.  1984.  *Hydraulics of Sediment Transport,* Water Resources Publications, Littleton, CO.

**Greenwood and Davis  1984**
Greenwood, B., and Davis, R. A., Jr.  1984.  *Hydrodynamics and Sedimentation in Wave-Dominated Coastal Environments*, Developments in Sedimentology 39, Elsevier, Amsterdam, the Netherlands (reprinted from *Marine Geology,* Vol 60, Nos. 1-4).

**Guza and Inman  1975**
Guza, R. T., and Inman, D. L.  1975.  "Edge Waves and Beach Cusps," *Journal of Geophysical Research*, Vol 80, pp 2997-3012.

**Hallermeier  1977**
Hallermeier, R. J.  1977.  "Calculating a Yearly Depth Limit to the Active Beach Profile,"  Technical Paper TP 77-9, Coastal Engineering Research Center, U.S. Army Engineer Waterways Experiment Station, Vicksburg, MS.

**Hallermeier  1978**
Hallermeier R. J.  1978.  "Uses for a Calculated Limit Depth to Beach Erosion," *Proceedings of the 16th Coastal Engineering Conference,* American Society of Civil Engineers, New York, pp 1493-1512.

**Hallermeier  1981a**
Hallermeier, R. J.  1981a.  "A Profile Zonation for Seasonal Sand Beaches from Wave Climate," *Coastal Engineering*, Vol 4, No. 3, pp 253-277.

**Hallermeier  1981b**
Hallermeier, R. J.  1981b.  "Terminal Settling Velocity of Commonly Occurring Sand Grains," *Sedimentology*, Vol 28, No. 6, pp 859-865.

**Hallermeier  1981c**
Hallermeier, R. J.  1981c. "Seaward Limit of Significant Sand Transport by Waves:  An Annual Zonation for Seasonal Profiles," Coastal Engineering Technical Aide CETA 81-2, Coastal Engineering Research Center, U.S. Army Engineer Waterways Experiment Station, Vicksburg, MS.

**Hands 1976**
Hands, E. B. 1976. "Observations of Barred Coastal Profiles Under the Influence of Rising Water Levels, Eastern Lake Michigan, 1967-71," Technical Report 76-1, Coastal Engineering Research Center, U.S. Army Engineer Waterways Experiment Station, Vicksburg, MS.

**Hands 1979**
Hands, E. B. 1979. "Changes in Rates of Shore Retreat, Lake Michigan 1967-76," Technical Paper No. 79-4, U.S. Army Engineer Waterways Experiment Station, Coastal Engineering Research Center, Vicksburg, MS.

**Hands 1980**
Hands, E. B. 1980. "Prediction of Shore Retreat and Nearshore Profile Adjustments to Rising Water Levels on the Great Lakes," Technical Paper No. 80-7, U.S. Army Engineer Waterways Experiment Station, Coastal Engineering Research Center, Vicksburg, MS.

**Hands 1983**
Hands, E. B. 1983. "The Great Lakes as a Test Model for Profile Response to Sea Level Changes," Chapter 8 in *Handbook of Coastal Processes and Erosion,* P. D. Komar, ed., CRC Press, Inc., Boca Raton, FL. (Reprinted in Miscellaneous Paper CERC-84-14, Coastal Engineering Research Center, U.S. Army Engineer Waterways Experiment Station, Vicksburg, MS.)

**Hanson and Kraus 1989**
Hanson, H., and Kraus, N. C. 1989. "GENESIS: Generalized Model for Simulating Shoreline Change; Report 1, Technical Reference, Technical Report CERC-89-19, U.S. Army Engineer Waterways Experiment Station, Vicksburg, MS.

**Harms 1969**
Harms, J. C. 1969. "Hydraulic Significance of Some Sand Ripples," *Bulletin of the Geological Society of America,* Vol 80, pp 363-396.

**Henkel 1970**
Henkel, D. J. 1970. "The Role of Waves in Causing Submarine Landslides," *Geotechnique,* Vol 20, pp 75-80.

**Horikawa 1988**
Horikawa, K., ed. 1988. *Nearshore Dynamics and Coastal Processes: Theory, Measurement and Predictive Models,* University of Tokyo Press, Tokyo, Japan.

**Houston 1995**
Houston, J. R. 1995. Beach Nourishment, *Shore and Beach,* Journal of the American Shore and Beach Preservation Association, Vol. 63, No. 1.

**Houston 1996a**
Houston, J. R. 1996a. The Economic Value of Beaches, *Proceedings, 1996 National Conference on Beach Preservation Technology*, St. Petersburg, Florida, Jan 24-26, 1996, pp. 271-280.

**Houston 1996b**
Houston, J. R. 1996b. International Tourism & U.S. Beaches, *Shore and Beach*, Journal of the American Shore and Beach Preservation Association, Vol. 64, No. 2, pp. 3-4.

**Hubbard, Oertel, and Nummedal 1979**

Hubbard, D. K., Oertel, G., and Nummedal, D. 1979. "The Role of Waves and Tidal Currents in the Development of Tidal-Inlet Sedimentary Structures and Sand Body Geometry: Examples from North Carolina, South Carolina, and Georgia," *Journal of Sedimentary Petrology*, Vol 49, No. 4, pp 1073-1092.

**Hume and Herdendorf 1988**
Hume, T. M., and Herdendorf, C. E. 1988. "A Geomorphic Classification of Estuaries and Its Application to Coastal Resource Management - a New Zealand Example," *Journal of Ocean and Shoreline Management*, Vol 11, pp 249-274.

**Jopling 1966**
Jopling, A. V. 1966. "Some Principles and Techniques Used in Reconstructing the Hydraulic Parameters of a Paleoflow Regime," *Journal of Sedimentary Petrology*, Vol 36, pp 5-49.

**Keulegan 1967**
Keulegan, G. H. 1967. "Tidal Flow in Entrances: Water-level Fluctuations of Basins in Communication with Seas," Technical Bulletin 14, Committee on Tidal Hydraulics, U.S. Army Corps of Engineers, Washington, DC.

**Kieslich 1977**
Kieslich, J. M. 1977. "A Case History of Port Mansfield Channel, Texas," General Investigation of Tidal Inlets Report 12, U.S. Army Engineer Waterways Experiment Station, Vicksburg, MS.

**King 1972**
King, C. A. M. 1972. Beacher and Coasts, 2nd ed., Edward Arnold, London, UK.

**Kinsman 1965**
Kinsman, B. 1965. *Wind Waves,* Prentice-Hall, Englewood Cliffs, NJ.

**Komar 1998**
Komar, P. D. 1998. *Beach Processes and Sedimentation,* 2nd ed., Prentice-Hall, Upper Saddle River, NJ.

**Komar et al. 1991**
Komar, P. D., et al. (Scientific Committee on Ocean Research (SCOR) Working Group 89.) 1991. "The Response of Beaches to Sea-Level Changes: A Review of Predictive Models," *Journal of Coastal Research*, Vol 7, No. 3, pp 895-921.

**Kraus and Harikai 1983**
Kraus, N. C., and Harikai, S. 1983. "Numerical Model of the Shoreline Change at Oarai Beach," *Coastal Engineering*, Vol 7, No. 1, pp 1-28.

**Kraus, Gorman, and Pope (eds.) 1994)**
Kraus, N. C., Gorman, L. T., and Pope, J. (eds.) 1994. "Kings Bay Coastal and Estuarine Physical Monitoring and Evaluation Program: Coastal Studies, Volume I, Main Text and Appendix A," Technical Report CERC 94-9, U.S. Army Engineer Waterways Experiment Station, Coastal Engineering Research Center, Vicksburg, MS.

**Kraus, Gorman, and Pope (eds.) 1995**
Kraus, N. C., Gorman, L. T., and Pope, J. (eds.) 1995. "Kings Bay Coastal and Estuarine Physical Monitoring and Evaluation Program: Coastal Studies, Volume II, Appendices B-G," Technical Report CERC 94-9, U.S. Army Engineer Waterways Experiment Station, Coastal Engineering Research Center, Vicksburg, MS.

**Larson 1991**
Larson, M. 1991. "Equilibrium Profile of a Beach with Varying Grain Size," *Coastal Sediments '91,* American Society of Civil Engineers, New York, NY, pp 905-919.

**Larson and Kraus 1989a**
Larson, M., and Kraus, N. C. 1989a. "SBEACH: Numerical Model for Simulating Storm-Induced Beach Change; Report 1, Empirical Foundation and Model Development." Technical Report CERC-89-9, U.S. Army Engineer Waterways Experiment Station, Vicksburg, MS.

**Leeder 1982**
Leeder, M. R. 1982. *Sedimentology: Process and Product,* George Allen and Unwin, London, UK.

**Le Méhauté 1976**
Le Méhauté, B. 1976. *An Introduction to Hydrodynamics and Water Waves,* Springer-Verlag, New York, NY.

**Lewis 1984**
Lewis, D. W. 1984. *Practical Sedimentology,* Hutchinson Ross, Stroudsburg, PA.

**Middleton 1965**
Middleton, G. V., Compiler. 1965. *Primary Sedimentary Structures and Their Hydrodynamic Interpretation,* Society of Economic Paleontologists and Mineralogists Special Publication No. 12, Tulsa, OK.

**Middleton and Southard 1984**
Middleton, G. V., and Southard, J. B. 1984. "Mechanics of Sediment Transport," Society for Sedimentary Geology (SEPM), Short Course No. 3, Tulsa, OK.

**Moore 1982**
Moore, B. D. 1982. "Beach Profile Evolution in Response to Changes in Water Level and Wave Height," M. S. thesis, University of Delaware, Newark, DE.

**Morang 1992a**
Morang, A. 1992a. "A Study of Geologic and Hydraulic Processes at East Pass, Destin, Florida," (in two volumes), Technical Report CERC-92-5, U.S. Army Engineer Waterways Experiment Station, Vicksburg, MS.

**Morang 1992b**
Morang, A. 1992b. "Inlet Migration and Geologic Processes at East Pass, Florida," *Journal of Coastal Research,* Vol 8, No. 2, pp 457-481.

**Morang 1998**
Morang, A. 1998. "Atlantic Coast of New York Monitoring Project; Report 1, Analysis of Beach Profiles, 1995-1996, Report prepared for U.S. Army Engineer District, New York, U.S. Army Engineer Waterways Experiment Station, Vicksburg, MS.

**Morang 1999**
Morang, A. 1999. Coastal Inlets Research Program, Shinnecock Inlet, New York, Site Investigation, Report 1, Morphology and Historical Behavior, Technical Report CHL-98-32, U.S. Army Engineer Waterways Experiment Station, Vicksburg, MS.

**Nairn, Zuzek, Morang, and Parson 1997**
Nairn, R. B., Zuzek, P., Morang, A., and Parson, L. 1997. "Effectiveness of Beach Nourishment on Cohesive Shores, St. Joseph, Lake Michigan," Technical Report CHL-97-15, U.S. Army Engineer Waterways Experiment Station, Vicksburg, MS.

**Nersesian and Bocamazo 1992**
Nersesian, G. K., and Bocamazo, L. M. 1992. Design and construction of Shinnecock Inlet, New York. *Coastal Engineering Practice '92,* American Society of Civil Engineers, New York, pp. 554-570.

**Niedoroda, Swift, and Hopkins 1985**
Niedoroda, A. W., Swift, D. J. P., and Hopkins, T. S. 1985. "The Shoreface," *Coastal Sedimentary Environments,* R. A. Davis, Jr., ed., 2nd ed., Springer-Verlag, New York.

**Nummedal and Fischer 1978**
Nummedal, D., and Fischer, I. A. 1978. "Process-Response Models for Depositional Shorelines: The German and the Georgia Bights," *Proceedings of the Sixteenth Conference on Coastal Engineering*, American Society of Civil Engineers, New York, pp 1215-1231.

**Nummedal and Humphries 1978**
Nummedal, D., and Humphries, S. M. 1978. "Hydraulics and Dynamics of North Inlet, South Carolina, 1975-76," General Investigation of Tidal Inlets Report 16, U.S. Army Engineer Waterways Experiment Station, Vicksburg, MS.

**Nummedal and Penland 1981**
Nummedal, D., and Penland, S. 1981. "Sediment Dispersal in Norderneyer Seegat, West Germany," *Holocene Marine Sedimentation in the North Sea Basin*, S. D. Nio, R. T. E. Schuttenhelm, and C. E. van Weering, eds., International Association of Sedimentologists Special Publication No. 5, pp 187-210.

**O'Brien 1931**
O'Brien, M. P. 1931. "Estuary Tidal Prisms Related to Entrance Areas," *Civil Engineering*, Vol 1, pp 738-739.

**O'Brien 1976**
O'Brien, M. P. 1976. "Notes on Tidal Inlets on Sandy Shores," General Investigation of Tidal Inlets Report 5, U.S. Army Engineer Waterways Experiment Station, Vicksburg, MS.

**Oertel 1982**
Oertel, G. F. 1982. "Inlets, Marine-Lagoonal and Marine Fluvial," *The Encyclopedia of Beaches and Coastal Environments*, M. L. Schwartz, ed., Hutchinson Ross Publishing Company, Straudsburg, PA, p 489.

**Oertel 1988**
Oertel, G. F. 1988. "Processes of Sediment Exchange Between Tidal Inlets, Ebb Deltas, and Barrier Islands," *Hydrodynamics and Sediment Dynamics of Tidal Inlets*, Lecture Notes on Coastal and Estuarine Studies, Vol 29, D. G. Aubrey and L. Weishar, eds., Springer-Verlag, New York, pp 297-318.

**Pilkey 1993**
Pilkey, O. H. 1993. "Can We Predict the Behavior of Sand: In a Time and Volume Framework of Use to Humankind?" *Journal of Coastal Research*, Vol 9, No. 1, pp iii-iv.

**Pilkey, Young, Riggs, Smith, Wu, and Pilkey 1993**
Pilkey, O. H., Young, R. S., Riggs, S. R., Smith, A. W. S., Wu, H., and Pilkey, W. D. 1993. "The Concept of Shoreface Profile of Equilibrium: A Critical Review," *Journal of Coastal Research*, Vol 9, No. 1, pp 225-278.

**Pratt and Stauble 2001**
Pratt, T., and Stauble, D. 2001. Coastal Inlets Research Program, Shinnecock Inlet, New York, Site Investigation, Report 3, Selected Field Data Report for 1997, 1998, and 1998 Velocity and Sediment Surveys. Technical Report CHL-98-32, U.S. Army Corps of Engineers, Engineer Research and Development Center, Vicksburg, MS.

**Price 1968**
Price, W. A. 1968. "Tidal Inlets," *The Encyclopedia of Geomorphology*, Encyclopedia of Earth Sciences Series, Vol III, R. W. Fairbridge, ed., Reinhold Book Corp., NY, pp 1152-1155.

**Price and Parker 1979**
Price, W. A., and Parker, R. H. 1979. "Origins of Permanent Inlets Separating Barrier Islands and Influence of Drowned Valleys on Tidal Records Along the Gulf Coast of Texas," *Transactions Gulf Coast Association of Geological Societies*, Vol 29, pp 371-385.

**Prior and Coleman 1979**
Prior, D. B., and Coleman, J. M. 1979. "Submarine Landslides - Geometry and Nomenclature," *Zeitschrift für Geomorphology,* Vol 23, No. 4, pp 415-426.

**Prior and Coleman 1980**
Prior, D. B., and Coleman, J. M. 1980. "Sonograph Mosaics of Submarine Slope Instabilities, Mississippi River Delta," *Marine Geology,* Vol 36, pp 227-239.

**Reineck and Singh 1980**
Reineck, H. E., and Singh, I. B. 1980. *Depositional Sedimentary Environments,* 2nd ed., Springer-Verlag, Berlin, Germany.

**Resio and Hands 1994**
Resio, D. T., and Hands, E. B. 1994. "Understanding and Interpreting Seabed Drifter (SBD) Data," Technical Report DRP-94-1, U.S. Army Engineer Waterways Experiment Station, Vicksburg, MS.

**Russell 1967**
Russell, R. 1967. "Origin of estuaries," *Estuaries,* G. H. Lauff, ed., American Association for the Advancement of Science, Publication 83, Washington, DC., 93-99.

**Sha 1990**
Sha, L. P. 1990. *Sedimentological Studies of the Ebb-tidal Deltas Along the West Frisian Islands, the Netherlands*, Geologica Ultraiectina No. 64, Instituut voor Aardwetenschappen der Rijks-universiteit te Utrecht, Ultrecht, The Netherlands (in English).

**Shepard 1950**
Shepard, F. P. 1950. "Longshore Bars and Longshore Troughs." Technical Memorandum 41, Beach Erosion Board, U.S. Army Corps of Engineers, Washington, DC.

**Short 1991**
Short, A. D. 1991. "Macro-meso Tidal Beach Morphodynamics - An Overview," *Journal of Coastal Research*, Vol 7, No. 2, pp 417-436.

**Stauble, Da Costa, Monroe, and Bhogal 1988**
Stauble, D. K., Da Costa, S. L., Monroe, K. L., and Bhogal, V. K. 1988. "Inlet Flood Tidal Delta Development Through Sediment Transport Processes," *Hydrodynamics and Sediment Dynamics of Tidal Inlets*, Lecture Notes on Coastal and Estuarine Studies, D. G. Aubry and L. Weishar, eds., Vol 29, Springer-Verlag, New York, NY, pp 319-347.

**Stauble, Garcia, Kraus, Grosskopf, and Bass 1993**
Stauble, D. K., Garcia, A. W., Kraus, N. C., Grosskopf, W. G., and Bass, G. P. 1993. "Beach Nourishment Project Response and Design Evaluation, Ocean City, Maryland," Technical Report CERC-93-13, U.S. Army Engineer Waterways Experiment Station, Vicksburg, MS.

**Suter and Berryhill 1985**
Suter, J. R., and Berryhill, H. L., Jr. 1985. "Late Quaternary Shelf-Margin Deltas, Northwest Gulf of Mexico," *Bulletin of the American Association of Petroleum Geologists,* Vol 69, No. 1, pp 77-91.

**Swift 1976**
Swift, D. J. P. 1976. "Coastal Sedimentation," *Marine Sediment Transport and Environmental Management*, D. J. Stanley and D. J. P. Swift, eds., John Wiley and Sons, New York, NY, pp 255-310.

**The Times Atlas of the World 1980**
*The Times Atlas of the World*. 1980. Comprehensive edition, Times Books, New York.

**U.S. Army Corps of Engineers 1958**
U.S. Army Corps of Engineers. 1958. Moriches and Shinnecock Inlets, Long Island, New York, "Survey Report, U.S. Army Engineer District, New York (original report dated September 1957, revised 11 July 1958).

**Walton and Adams 1976**
Walton, T. L., Jr., and Adams, W. D. 1976. "Capacity of Inlet Outer Bars to Store Sand," *Proceedings of the Fifteenth Coastal Engineering Conference*, July 11-17, Honolulu, HI, American Society of Civil Engineers, New York, pp 1919-1937.

**Williams, Morang, and Lillycrop 1998**
Williams, G. L., Morang, A., and Lillycrop, L. 1998. "Shinnecock Inlet, New York, Site Investigation; Report 2, Evaluation of Sand Bypass Options," Technical Report CHL-98-3, U.S. Army Engineer Waterways Experiment Station, Vicksburg, MS.

**Wright 1981**
Wright, L. D. 1981. "Nearshore Tidal Currents and Sand Transport in a Macrotidal Environment," *Geomarine Letters*, Vol 1, pp 173-179.

**Wright 1985**
Wright, L. D. 1985. "River Deltas," *Coastal Sedimentary Environments,* 2nd ed., R. A. Davis, ed., Springer-Verlag, New York, pp 1-76.

**Wright and Coleman 1972**
Wright, L. D., and Coleman, J. M. 1972. "River Delta Morphology: Wave Climate and the Role of the Subaqueous Profile," *Science*, Vol 176, pp 282-284.

**Wright and Coleman 1973**
Wright, L. D., and Coleman, J. M. 1973. "Variations in Morphology of Major River Deltas as Functions of Ocean Wave and River Discharge Regimes," *American Association of Petroleum Geologists Bulletin*, Vol 57, No. 2, pp 370-398.

**Wright and Coleman 1975**
Wright, L. D., and Coleman, J. M. 1975. "Mississippi River Mouth Processes: Effluent Dynamics and Morphologic Development," *Journal of Geology,* Vol 82, pp 751-778.

**Wright and Short 1984**
Wright, L. D., and Short, A. D. 1984. "Morphodynamic Variability of Surf Zones and Beaches: A Synthesis," *Marine Geology*, Vol 56, pp 93-118.

**Wright and Sonu 1975**
Wright, L. D., and Sonu, C. J. 1975. "Processes of Sediment Transport and Tidal Delta Development in a Stratified Tidal Inlet," *Estuarine Research*, Vol 2, L. E. Cronin, ed., Academic Press, New York, pp 63-76.

**Wright, Boon, Kim, and List 1991**
Wright, L. D., Boon, J. D., Kim, S. C., and List, J. H. 1991. "Modes of Cross-Shore Sediment Transport on the Shoreface of the Middle Atlantic Bight," *Marine Geology*, Vol 96, pp 19-51.

**Wright, Sonu, and Kielhorn 1972**
Wright, L. D., Sonu, C. J., and Kielhorn, W. V. 1972. "Water-Mass Stratification and Bed Form Characteristics in East Pass, Destin, Florida," *Marine Geology*, Vol 12, pp 43-58.

**Wright, Xu, and Madsen 1994**
Wright, L. D., Xu, J. P., and Madsen, O. S. 1994. "Across-shelf Benthic Transports on the Inner Shelf of the Middle Atlantic Bight During the 'Halloween Storm' of 1991," *Marine Geology*, Vol 118, No. 1/2, pp 61-77.

**Young and Hale 1998**
Young, C., and Hale, L. 1998. Coastal Management: Insurance for the Coastal Zone, *Maritimes*, Vol 40m, No. 1, pp. 17-19.

**Zenkovich 1967**
Zenkovich, V. P. 1967. "Submarine Sandbars and Related Formations," *Processes of Coastal Development*, J. A. Steers, ed., Oliver and Boyd, Ltd., New York, pp 219-236.

## IV-3-7. Definition of Symbols

| | |
|---|---|
| $\beta$ | Gradient of the beach and surf zone |
| $\varepsilon$ | Surf-scaling parameter (Equation IV-3-4) [dimensionless] |
| $\rho_f$ | Mass density of fresh water ( = 1,000kg/m$^3$ or 1.94 slugs/ft$^3$) [force-time$^2$/length$^4$] |
| $\rho_s$ | Mass density of salt water ( = 1,025 kg/m$^3$ or 2.0 slugs/ft$^3$) [force-time$^2$/length$^4$] |
| $\omega$ | Wave angular or radian frequency (= $2\pi/T$) [time$^{-1}$] |
| $\Omega$ | Modal state of the beach (Equation IV-3-3) [dimensionless] |
| $A$ | Sediment scale or equilibrium profile parameter or profile shape parameter (Table III-3-3) [length$^{1/3}$] |
| $a_b$ | Breaker amplitude [length] |
| $B$ | Berm height of the eroded area [length] |
| $d_\ell$ | Annual depth of closure below mean low water (Equation IV-3-9) [length] |
| $F'$ | Froude number |
| $g$ | Gravitational acceleration (32.17 ft/sec$^2$, 9.807m/sec$^2$) [length/time$^2$] |
| $h$ | Equilibrium beach profile depth (Equation IV-3-7) [length] |
| $H_*$ | Water depth [length] |
| $H_b$ | Wave height at breaking [length] |
| $H_e$ | Non-breaking significant wave height that is exceeded 12 hr per year [length] |
| $H_x$ | Wave height of the $x$-year return period [time] |
| $h'$ | Depth of density interface [length] |
| $L_*$ | Cross-shore distance to the water depth $H_*$ [length] |
| $R$ | Shoreline retreat (Equation IV-3-5) [length] |
| $S$ | Increase in sea level [length] |
| $T$ | Wave period [time] |
| $T_e$ | Wave period associated with $H_e$ [time] |
| $U$ | Mean outflow velocity of upper layer (in case of stratified flow) [length/time] |
| $\overline{w_s}$ | Sediment fall velocity [length/time] |
| $x$ | Retreat of the profile, Bruun Rule (Equation IV-3-6) [length] |
| $X$ | Horizontal distance of responding profile {Equation IV-3-6 and Figure IV-3-40) [length] |

| | |
|---|---|
| $y$ | Equilibrium beach profile distance offshore (Equation IV-3-7) [length] |
| $z$ | Change in water level [length] |
| $Z$ | Closure depth (Equation IV-3-10) [length] |
| $Z$ | Vertical distance of responding profile {Equation IV-3-6 and Figure IV-3-40) [length] |

## IV-3-8. Acknowledgments

Authors of Chapter IV-3, "Coastal Morphodynamics:"

Andrew Morang, Ph.D., Coastal and Hydraulics Laboratory (CHL), Engineer Research and Development Center, Vicksburg, Mississippi.

Larry E. Parson, U.S. Army Engineer District, Mobile, Mobile, Alabama.

Reviewers:

Joan Pope, CHL

William E. Birkemeier, CHL

Stephan A. Chesser, U.S. Army Engineer District, Portland, Portland, Oregon.

Ronald L. Erickson, U.S. Army Engineer District, Detroit, Detroit, Michigan.

Edward B. Hands, CHL (retired).

Edward Meisburger, CHL (retired).

Joan Pope, CHL

John F. C. Sanda, Headquarters, U.S. Army Corps of Engineers, Washington, DC., (retired).

Orson P. Smith, Ph.D., U.S. Army Engineer District, Alaska, Anchorage, Alaska, (retired).